Responding to "Routine" Emergencies Workbook

Responding to "Routine" Emergencies Workbook

Frank C. Montagna

Fire Engineering

PennWell®
MEDIA FOR STRATEGIC MARKETS   SINCE 1910

Copyright © 2006 by PennWell Corporation
1421 South Sheridan Road
Tulsa, OK 74112-6600 USA
800-752-9764
www.pennwellbooks.com
www.pennwell.com

Director: Mary McGee
Managing Editor: Steve Hill
Production/Operations Manager: Traci Huntsman
Production Manager: Robin Remaley
Assistant Editor: Amethyst Hensley
Production Assistant: Amanda Seiders
Book Designer: Clark Bell

Library of Congress Cataloging-in-Publication Data Available on Request

Montagna, Frank
Responding to "Routine" Emergencies Workbook

ISBN 1-59370-047-4
ISBN13 978-1-59370-047-8

Printed in the United States of America

4 5 6 7 8   17 16 15 14 13

Throughout my 35 years in the fire service, it has been my good fortune to have exceptionally knowledgable and generous mentors and friends both in the New York City Fire Department as well as in the fire service around the country, all of whom freely shared their knowledge and experience with me.

I would like to dedicate this book to everyone who shared their knowledge with me and told me "stuff." Without them I would not have been able to tell "stuff" to anyone else.

In fact, without their help, wisdom, and shared experience, I might not have survived.

Contents

How to Use This Book

- After reading a chapter in Responding to "Routine" Emergencies, the student should answer the questions in this book that relate to the chapter.

- Any wrong answers should be looked up in the textbook and reviewed. Page numbers have been included for each question. Simply refer back to the posted page number in the text.

- After reading the book, answering the questions, and reviewing the material for each wrong answer, read and answer the scenario.

Introduction

Since my book *Responding to "Routine" Emergencies* was published in 1999, I have received numerous requests for a companion workbook to go along with it. In reply to these requests, I began to put monthly quizzes on my Web page http://www.chiefmontagna.com. They were well received and are still posted although I have not added to them since June 2002. At that time, I began working on this book.

The questions are broken down by chapter and are multiple choice, true or false/correct or incorrect, and short-answer questions. I also include a scenario for most chapters. There is only one scenario for Part II of the book, the carbon monoxide section. That makes a total of 9 scenarios in all. In addition, I suggest several drills for most chapters. These drills can be conducted in the firehouse or outdoors in your response area. Most of the drills will require that you do some preparation in order to give an effective drill. The effort involved in preparing for the drill will not only ensure a worthwhile training session but will also enhance your own knowledge of the topic and hopefully suggest additional avenues that you can drill on.

For those of you studying for a promotional exam, I have included a short section in this book in which I list some tips for taking multiple-choice tests. Many promotional tests include this type of question, and good test-taking techniques can improve your score. There are a number of Web sites that have test-taking tips available and many books have been written on the topic. The few tips that I have included are basic but can have a serious impact on your score if ignored. Good luck on your test.

I hope that you find this book useful and that it helps keep you and your firefighters safe. Remember: There are no *routine* responses.

Frank Montagna

January 2005

Test-taking Tips for Multiple-choice Tests

- Read the entire question before you look at the answers. Sometimes an answer will be partially correct, or one of the later choices will be more correct.

    – For example:

Question:

Why do we read PennWell Books? Pick the most correct answer.

A. Because we want to learn all we can about firefighting.

B. Answer A plus because they are written by knowledgeable firefighters.

C. Answer B plus because the material in them is accurate.

D. Answer C plus because the knowledge in them can save our lives.

The answer is obviously D. Not because A, B, or C are wrong, but because D is the most correct.

- If you can, decide on what the answer should be before you look at the available choices.

- Some tests only count your correct answers and charge no penalty for questions not answered. Don't guess on these tests. If you don't know the answer, don't answer the question. Most tests do penalize you for missing answers. On these tests, you should leave no question blank. If you don't know the answer, take your best guess.

- Skipping a question and leaving it blank can result in placing the next question in the wrong place. For example: You do not know the answer to question 21 so you decide to answer it later. You might inadvertently place the answer to question 22 in the blank space provided for question 21. Every answer thereafter will be in the wrong place. To avoid that potential disaster, take your best guess and enter that answer for question 21. Then mark the question to remind you to go back and review it later after you have finished all the other questions.

- Try to increase your odds of guessing correctly by eliminating answers that you know are wrong. If a multiple-choice question has 4 choices, you have a 1 in 4 chance of guessing correctly. If you can eliminate 2 of the possible answers, then your odds have been improved to 50-50 or a 1 in 2 chance of getting it right.

- Don't change your answer without a good reason. Did you misread the question or answer the first time? OK, change your answer. Did you get additional information from another question in the test? Again OK, change it if you think it is wrong. Remember, answers that you change have a higher probability of being wrong.

- The words *always* or *never* should put you on your guard. They often are found in wrong answers. How many things in firefighting are *always* or *never* true?

Acronyms and Abbreviations

AIM (company name)

BLEVE (boiling-liquid expanding-vapor explosion)

CGI (combustible gas indicator)

CNG (compressed natural gas)

CO (carbon monoxide)

COHb (carboxyhemoglobin)

CPSC (Consumer Products Safety Commission)

EPA (Environmental Protection Agency)

FD (fire department)

FDNY (Fire Department of New York)

ft (feet)

gpm (gallons per minute)

HVAC (heating, ventilation, and air conditioning)

IAS (International Approval Services)

IAFC (International Association of Fire Chiefs)

hr (hour)

in. (inch)

min (minute or minutes)

NFPA (National Fire Protection Association)

OOS (out of service)

OSHA (Occupational Safety and Health Administration)

PCB (polychlorinatedbiphenyl)

PCBs (polychlorinatedbiphenyls)

PD (police department)

ppm (parts per million)

PPV (positive pressure venting)

PVC (polyvinylchloride)

SCBA (self-contained breathing apparatus)

T&P (temperature–pressure)

UL 2034 (CPSC standard)

Part One:
Common Calls

Electrical Emergencies

1

Questions

1. What is contained in the insulation of some wires, that when burned, contributes hydrogen chloride gas to the smoke?

2. What carcinogen is contained in some transformer coolant oils?

3. What would be a positive check for a defective ballast?

4. You should check the ceiling space above an overheating fluorescent fixture for possible fire extension. It is possible for an overheated ballast to ignite combustible fiberboard tiles. True or false?

5. Halogen lamps are a fire hazard. They can reach temperatures as high as __?__. (Fill in the blank.)

6. **You are called to a home because the occupant saw sparks coming from a light fixture. Which of the following statements about the actions that you should take are *false*?**

   A.   Leave the wiring safe by capping or taping leads that have been exposed by your examination.

   B.   Tripping the appropriate breaker or pulling the fuse will make the wires safe to handle.

   C.   If the problem is widespread, you may have to kill the power to the entire building.

   D.   Once the problem is located and made safe, restore the power to the building.

7. **You are called to a home for an electrical odor and you find that the metal clad cable on one circuit is heating dangerously. What hazard to operating firefighters remains even after removing the power to that cable?**

8. **Pick the *incorrect* statement.**

   A.   When examining for fire extension at an electrical fire, pay special attention to wires that pass through or lie on studs or beams.

   B.   Try to determine where the involved wiring is coming from and where it is going to.

   C.   Open an examination hole when the wall surface is hot.

   D.   Scanning the walls with a thermal imaging camera can reveal hidden pockets of fire and can identify overheated wires.

   E.   All statements are incorrect.

9. **Opening the main switch removes power from a residential building making the interior wiring safe to handle. True or false?**

10. Buildings that house heavy users of electricity and are supplied with 460V services will have the electric service in a locked room with a posted warning sign. Which of the following statements about these rooms are true?

A. Any contact with the exposed wire or bus found in these rooms will cause death.

B. A safety device will trip and cut off power if you contact the bus.

C. These rooms must be immediately but cautiously entered if an alarm inside is sounding.

D. If fire or smoke is extending from the locked room, then control extending fire, evacuate the area and ventilate while awaiting the utility personnel. Do not enter the room or use water inside the room.

11. Which of the following statements are accurate?

A. If an electrical problem in a building requires that the service be disconnected, any firefighter can remove the meter.

B. Lineman's gloves must be regularly inspected and tested.

C. Oil leaking from a faulty or damaged transformer may contain carcinogenic polychlorinatedbiphenyl (PCB) oil.

D. A downed TV cable wire is always safe to handle.

E. Once the power has been removed from a downed power line, it is safe .

12. You must learn to recognize and identify manhole covers and associate them with their proper utility and purpose. What are two ways manhole covers can be identified?

13. How can you distinguish between an electrical manhole containing service wiring and a manhole containing a transformer?

14. Why should you not stand over or near the grating covering a smoking underground transformer?

15. Transformers can heat because of overload, typically during times of great demand. What hazard exists when a transformer overheats?

16. The gases given off by a burning manhole are not only noxious and deadly but flammable and explosive. If a smoking manhole stops smoking, it is safe to say that the explosion danger is over. True or false?

17. Manhole covers can weigh more than __A__ pounds and can be blown __B__ feet or more into the air when the gaseous products of combustion in the manhole ignite explosively. Fill in the blanks.

18. Electrical manholes can supply power to streetlights and traffic lights. What danger does this pose to firefighters and civilians in the area of a manhole fire?

19. At electrical manhole fires, consider the danger area to include anything that has a connection to the manhole. At a minimum, locate and identify the __?__ on all sides of the trouble hole and keep people and cars away from it. Fill in the blank.

20. When doing a visual survey at a manhole fire, you must identify the obviously involved manholes as well as other manholes that could become involved. Why check the overhead wires?

21. You can measure the danger of a manhole incident by the amount and color of smoke that you see. True or false?

22. Which of the following statements about manhole fires are true?

    A. Check only the buildings directly opposite the manholes for fire extension and carbon monoxide (CO) infiltration.

    B. The problem in the manhole can cause a building's electric service to heat up and ignite nearby combustibles.

    C. Combustible and toxic gases can seep into buildings through underground ductwork.

    D. Deadly carbon dioxide gas can seep into the building from a burning manhole.

23. The appropriate way to flood a manhole is to approach no closer than 5 ft from the burning manhole and operate a low-pressure fog nozzle directly into the opening. True or false?

24. Once utility company personnel respond to an incident, they are responsible for the mitigation of the incident and the safety of all the people involved. True or false?

Topics for Drill

1. A. Name five hazards that may be present at a smoking manhole incident.
    B. What can be done to protect firefighters from each hazard?

2. What type of manholes do you have in your response area and what hazard do they pose to you? Get a photo of each type and use them at your next drill to familiarize firefighters with their hazards.

3. A. Name five sources of an electrical odor in a structure.
    B. Describe how each occurs and what responding firefighters must do to mitigate the incident.

4. Describe an odor of smoke incident that you responded to that could have turned deadly if you did not find the reason for the odor. Use this incident at drill to point out that even routine odor of smoke incidents can be hazardous to us or to those we have sworn to protect.

5. What are six precautions that you should take at a downed power line incident?

    A.   Draw a diagram of the typical overhead-wiring pole in your area and label all of the wires on the pole. Include primary, secondary, telephone, cable TV, and any other wiring present.

    B.   Describe what hazard each wire can pose to firefighters. Don't forget that a downed wire can charge a wire that would normally be considered safe.

    C.   Take a photo of the pole you used as your model and use it and the diagram at drill to point out the hazards present at downed wire emergencies.

Scenario: Manhole Emergency

Assume that you are a captain in charge of a ladder company. In addition to yourself, you have three firefighters working with you. You have responded to a report of smoke coming from a manhole in the street. As you pull up to the scene of the reported smoking manhole, you notice that you are in an area where electric service is provided to buildings by a combination of overhead and underground wires. Black smoke is pushing under pressure out of a grated rectangular manhole. The streetlights and the lights of some of the homes on the block are flickering and a number of residents are in the street looking at the smoking manhole and talking excitedly among themselves. Your response is one engine with three firefighters and an officer, your truck, and a chief. The chief carries a combustible gas indicator, and the truck carries a CO meter.

Scenario Questions

1. From a safe distance, you examine the manhole and observe that it is rectangular in shape and is covered by six grated panels. What does this tell you about the content of the manhole? What hazards should it alert you to?

2. As you approach the scene, what safety precautions should you consider when placing your apparatus?

3. What actions should you take?

4. One of your firefighters tells you that the lights in a nearby house are blinking. He asks you if you want him to kill the electricity to the home by opening the main shutoff or by pulling the meter. What do you tell him and why?

5. What help will you need?

6. Your first due engine has stretched a hose to the smoking transformer vault and is getting ready to put water into it to extinguish the fire. Do you agree with this tactic? Why or why not?

Answers

1. Polyvinylchloride (PVC)
   ⮔ *See book page 6*

2. Polychlorinatedbiphenyls (PCBs)
   ⮔ *See book pages 22, 30*

3. Feel the ballast. If it is so hot that you can't keep your hand on it, it is defective.
   ⮔ *See book page 7*

4. True.
   ⮔ *See book page 7*

5. 1,000° F.
   ⮔ *See book page 9*

6. A, B, and C are true.
   D is false. Restoring power is the responsibility of an electrician after he thoroughly checks the circuit for safety.
   ⮔ *See book page 14*

7. Once heated, the cable retains the heat. Use caution to avoid being burned.
   ⮔ *See book page 15*

8. E.
   ⮔ *See book page 15, 16*

9. True, unless power is pirated into the building from another source.
   ⮔ *See book page 17*

10. A and D are true.

    B is false. There is no safety device.

    C is false. In no case should fire department personnel enter such a room until the utility personnel kill the power.

    ⊃ *See book page 18*

11. B and C are true.

    A. False. Removing the meter from a building is best left to utility workers. They are trained and equipped to perform this task.

    ⊃ *See book page 19*

    D. False. A charged power line may be in contact with the uncharged cable. The charged line can be several blocks away and out of sight.

    ⊃ *See book page 23*

    E. The wire must be grounded to remove the residual electric potential from the downed lines before it is safe.

    ⊃ *See book page 23*

12. Some are inscribed with identifying marks and some have distinctive shapes.

    ⊃ *See book page 30*

13. Underground transformers are vented by gratings. These vaults are usually 4×6 ft or greater in size and can be covered with as many as five gratings.

    ⊃ *See book page 30*

14. These transformers can fail explosively, and the explosive force will be vented up through the grating. It is possible for parts of the transformer to be blown up and out of the hole.

    ⊃ *See book page 32*

15. As the transformer overheats, the oils within it expand and can eventually cause failure of the transformer walls.
⮑ *See book page 32*

16. False. A burning manhole that appears to have gone dormant can unexpectedly explode.
⮑ *See book page 34*

17. A. 300.
    B. 60.
⮑ *See book page 34*

18. Traffic light control boxes and light pole access panels have been blown off of the poles more than a block away from a burning manhole. Carbon monoxide from the manhole can enter the poles and traffic boxes via underground ducts, and when ignited by a spark, can result in an explosion.
⮑ *See book page 35*

19. Identify the next manhole on all sides of the trouble hole.
⮑ *See book page 35*

20. Overhead wires could have a connection to the underground wires. If so, fire can spread up the pole to the overhead wires.
⮑ *See book page 35*

21. False. Conditions can change drastically in a short time. Manhole covers may be sealed rather than vented and as a result, no smoke will be visible. What appears to be an uninvolved manhole can suddenly blow its lid.
⮑ *See book page 37*

22. B and C are true.
    A. Check any building that might receive electricity from the manhole.
    D. It's deadly carbon monoxide gas.
⮑ *See book page 37*

23. False. Lay an open but 2 ½ in. or larger hose on the ground, as far away as practical from the burning hole and flow water into the hole. Another option is to open a nearby hydrant and allow the water from it to flow into the hole.
   See book page 39

24. False. The fire chief has ultimate responsibility for the incident and safety. He must seek solutions from the experts and then critically evaluate this information for adverse implications on those he is sworn to protect.
   See book page 40

Scenario Answers

1. The manhole probably holds a transformer. Transformers are typically housed in rectangular manholes and vented by grated panels. The transformer either steps high voltage down to low voltage or low voltage up to high voltage.

   • The transformer is cooled by a dielectric fluid that could contain PCBs, a carcinogen. When it is contained it presents no hazard. If it leaks out, it is a hazmat incident. Even the smoke could be contaminated.

   • This smoke will contain a noxious brew of toxic and explosive gases including CO.

   • It is possible (under the right conditions) for the transformer to violently fail, blowing flaming oil, and metal parts out of the hole.

   • The smoke can be filtering into anything connected to the underground vault, including nearby buildings, light poles, and other structures.

   • If smoke is seeping into structures, it could sicken—even kill—building occupants. If the CO it contains is found in high enough concentrations, the smoke could be ignited by a spark, resulting in an explosion.

2. You should know that the next manhole on either side of the trouble hole might at any time become involved. It might start to smoke, erupt in flames or even blow its lid into the air. Find the next manhole on either

side and be sure you park the apparatus a safe distance away from it. In addition, in some instances, the underground problem can extend up to the overhead wires, so, as a precaution, do not park under the overhead wires.

3. Upon arrival, you should set up an exclusion area and keep civilians, traffic, and firefighters out of this area.

    *   Consider the possibility that built-up gasses might ignite explosively, blowing the manhole cover into the air, or that the transformer might explode violently.

    *   The exclusion area should include anywhere a flying manhole or metal transformer part can land as well as the area that is engulfed in the smoke from the manhole.

    *   Check the nearby houses for CO and other toxic or combustible gases. If you find them in a building, check the next couple of buildings until you have defined the danger area. You must continually recheck all buildings until the incident is concluded. If you find elevated levels of CO, evacuate the residents to a safe area.

    *   Stretch a precautionary line to a safe area with enough hose to cover the threatened areas.

    *   Position all apparatus in a safe area.

4. You should tell him not to open the main shutoff and not to pull the meter. Combustible gases may be in the circuit box or could have filtered up into the meter pan and could ignite as a result of a spark when the switch is opened or the meter is pulled. The result could be an explosion in the box or meter. It is too dangerous to cut the electric if CO is present. The blinking lights are not an immediate hazard.

5. You will definitely need your electric utility to respond.

    *   If there are gas lines in the area, have the gas utility respond also, especially if you smell gas. The burning transformer and wires can affect nearby gas lines. These incidents have resulted in gas mains—particularly plastic mains—being damaged. The result is a gas main leak and possible gas ignition added to your problems.

- Do you have enough manpower on the scene to check and recheck surrounding buildings and to stop both traffic and civilians from entering the danger area? If not, get more units to the scene. Don't forget to keep your firefighters out of the danger area. Call for police help with traffic and civilians.

- Do you have enough CO meters and combustible gas indicators to quickly check surrounding properties? The utility will have meters and can help you monitor CO levels, but how long will it be before they respond? Call more meter-equipped fire department (FD) units to the scene if you need immediate help.

- Is anyone feeling ill from the smoke? Get an ambulance to respond. How many buses will you need?

- How many people were evacuated? What are you going to do with them? Do they need protection from the elements? Get a bus to respond or the Red Cross or some other local agency to respond and help. If there is a nearby school or other large building, you may be able to house the evacuees in the building to get them out of the weather. Force entry to the building if it is after hours.

6. **No. Stop them. You should not put water on a transformer vault fire until requested to do so by utility personnel.**

   - Adding water will push the smoke into underground conduit and possibly create a CO problem where none may yet exist.

   - It is possible for the water to cause an arc that may result in transformer failure, an explosion, and oil release.

   - The utility personnel will have to pump the water out before they can repair the transformer.

   - Some transformers are not meant to be submerged and flooding the vault they are in could release the dielectric oil. The oil might be PCB-contaminated.

   - If you decide to put water into the vault after being requested to do so by utility personnel, let water flow into the vault from an unmanned open butt or bounce water off of the street and into the vault from your hose line. Do not stand near the manhole and play your stream directly into the manhole. Always assume it is live and keep your firefighters out of a potential dangerous situation.

Home Heating Emergencies 2

Questions

1. When responding to a home heating emergency, we should try to repair the problem that has caused the emergency. True or false?

2. Which of the following statements are accurate?

    A. Fuel oil has a flash point between 100°F and 130°F.

    B. Contamination with a lower flash-point product like gasoline will lower the flash point of fuel oil.

    C. No. 2 fuel oil is light, has low viscosity, flows easily, and is used in home oil burners; while No. 3 fuel oil is used in large commercial burners, has high viscosity, and must be preheated before it flows easily.

    D. All of the above are accurate.

3. **A malfunctioning oil burner may result in pooled excess oil continuing to burn in the fire box after the ignition cycle shuts down. The result is the generation of large amounts of black smoke. Which of the following statements correctly describe this condition?**

   A. Incomplete combustion occurs because the fan has shut down and the fire is not receiving enough air.

   B. Excess carbon monoxide and other toxic gases are generated.

   C. This condition is known as pre-burn.

   D. This can not happen, because of the safeties found on oil burners.

4. **A flue that is too long can result in excessively cooled flue gases that may condense in the flue pipe. Which of the following hazards can result from this condition?**

   A. The resulting acidic condensate can cause the metal flue to deteriorate over time, resulting in perforation of the metal flue, which allows toxic gases to spill into the living space.

   B. The draft can fail and the toxic flue gases will be spilled into the home rather than exiting the flue as intended.

5. **Which of the following statements are true?**

   A. You can expect to find all of the following size fuel oil tanks inside of a private dwelling: 275 gallons, 550 gallons, and 1,080 gallons.

   B. There will only be one fuel tank in a private dwelling.

   C. In some locations, fuel oil tanks are required to be vaulted.

6. **Which of the following statements are accurate?**

   A. The fuel tank vent pipe lets air exit the tank as fuel enters it.

   B. Some vent pipes have a whistle alarm.

   C. If the vent pipe is blocked, the alarm will still sound.

   D. If the vent pipe is blocked, a pressure build-up will occur that could quickly result in tank rupture.

7. **Which of the following statements about oil burners are not true?**

    A. There will be only one fuel oil shutoff found either at the oil burner or at the fuel storage tank.

    B. There will be a fuel shutoff valve at the oil burner and possibly another at the tank.

    C. If you see a fill pipe on the exterior of the building and an oil tank inside, you can be sure the heating unit is oil-fired.

    D. Fuel oil need not be atomized before it is ignited.

8. **Safety devices allow the oil burner's fuel pump to function for no more than 90 seconds if no flame is detected. Even so, about eight ounces of fuel or more can be pumped into the combustion chamber before the safety shuts the burner down. This results in unburned fuel collecting in the combustion chamber. There is no particular hazard associated with this phenomenon. True or false?**

9. **If an oil burner's barometric draft damper door does not swing freely, what problem can develop in the heating system?**

10. **The oil burner's electrical system uses a step-up transformer that receives energy at 120V and produces energy as high as __?__ volts for the spark that ignites the oil. Pick the correct answer.**

    A. 1,000 to 2,000

    B. 5,000 to 8,000

    C. 10,000 to 12,000

    D. 15,000 to 18,000

11. **Which of the following statements are true?**

A. The thermostat controls the starting and the stopping of the ignition cycle in all heating systems.

B. The burner emergency shutoff is usually located at the door to the burner room or at the top of the basement stairs.

C. The oil burner shutoff switch often has a yellow switch plate.

D. Residential boilers operate at a pressure of 2 psi for hot water and 15 psi for steam.

12. **Which of the following statements are true about hot water heating systems?**

A. If a hot water heating system has no low-water cutoff, it can dry fire if the feed water is cut off.

B. This will cause the boiler to become extremely hot but will not result in any damage to the boiler.

C. When the thermostat calls for heat, the circulator will kick in and pump a few gallons of residual water into the hot boiler where it will be converted to steam.

D. The resultant temperature change may blow the boiler apart.

13. **Which of the following statements are true?**

A. The primary control protects the oil burner from failure of the ignition or flame.

B. Delayed ignition or puffback occurs when unburned atomized fuel is ignited at the end of a burner ignition cycle, resulting in a small explosion.

C. A puffback is accompanied with a low bang or thud accompanied by light wispy smoke.

D. The flue pipe may be dislodged in a puffback.

14. We respond to a report of black smoke in the oil burner room and find that the oil fire is out of the combustion chamber. Is this an emergency or a structural fire?

15. When afterfire occurs, thick black smoke will be seen exiting the chimney. If smoke has filled the basement, what can be assumed?

16. Which of the following actions should you take at an afterfire incident?

   A.   Shut off the burner.
   B.   Shut off the fuel.
   C.   Ventilate if necessary.
   D.   Extinguish any fire that has extended from the burner.

17. At an afterfire incident, if there is a danger of flames extending to the structure, what action may be required?

18. What oil burner emergency can cause the whole house to shake and has been likened to a train passing nearby?

19. To prevent the burner door from opening during pulsation, a homeowner might brace a 2×4 against the door. What danger does this pose?

20. **You are called to an oil burner incident and are met with heavy black smoke pouring from the basement windows and door. As the incident progresses, the smoke suddenly changes in the color from black to pearly white accompanied by an oily smell and taste. This change in color and oily smell is a warning of what potentially deadly hazard?**

21. **Which of the following might occur at a white ghost incident?**

    A.   An explosion.

    B.   Possible partial structural collapse.

    C.   Rapid fire extension.

    D.   The smoke will suddenly turn black and feel cool to the touch.

22. **Which of the following statements are accurate?**

    A.   You are in a smoky burner room and the smoke changes from black to pearly white. You should know that this means that the fire has been knocked down and all is well.

    B.   Oil burner smoke from a puffback is usually cool and easy to breathe. Self-contained breathing apparatus (SCBA) is not needed.

    C.   High heat emanating from the oil burner room, accompanied by thick black smoke, is an indication of a structural fire.

23. **Pick the correct answers.**

   A.   The temperature-pressure (T&P) valve located on top or side of a water heater tank is designed to vent excess pressure when the water temperature reaches 210°F or 150 psi.

   B.   If the T&P valve on a water heater fails to operate, other safeties will prevent a pressure build up.

   C.   Steam or hot water discharging from the T&P valve or steam rather than hot water discharging from the hot water faucet are the symptoms of a defective low water cut off.

   D.   A cherry red water heater and the smell of hot metal are the signs of a defective T&P valve.

24. **Which of the following actions should be taken when responding to an overheated boiler or hot water heater?**

   A.   Shut down the fuel and electrical supply.

   B.   Allow cool water to flow into the heater to cool it down.

25. **Which of the following statements are true?**

   A.   If the pilot light of a hot water heater is extinguished and the safety shutoff valve sticks open or is blocked by debris, the gas will continue to flow and leak into the atmosphere.

   B.   Delayed ignition in a gas hot water heater can result in a flame being projected beyond the combustion chamber.

Topics for Drill

Create a list of sites in your district that have large commercial oil burners. Visit each site and ascertain if they use heavy oil that must be preheated before use. Then locate the gas shutoff. Also compile a list of occupancies that use a dual-fueled burner and locate their gas shutoffs.

1. What are the general precautions we must take when responding to oil burner emergencies?
   - For each answer, explain the hazards involved.

2. What are the general tactics for fighting an oil burner fire?
   - Describe why or when each tactic is necessary.

3. Go to your firehouse's furnace and or water heater.
   A. Point out the various parts of the furnace or water heater.
   B. Point out the various safeties and explain what they do and the result of their malfunction.
   C. Locate the electric and fuel shutoffs. Explain why you must shut them off.
   D. If you use fuel oil, find out what grade. How much fuel oil does your tank hold? What action would you take if the tank ruptured?

Scenario: Oil Burner Emergency

You are the captain of a truck company responding with four firefighters. Your department staffs engine companies with three firefighters and an officer. It is the first cold night of the fall heating season, and you are called to respond to a smoke condition in the hallways of a 3-story woodframe house at 0200 hours. The building houses one family per floor for a total of three families.

Your response is one engine, a truck, and the chief. The chief carries a CO detector. Upon arrival at the scene, you smell the distinctive odor of burning fuel oil and see light smoke haze flowing out of the opened cellar windows and the opened first floor entrance door. There is a woman waving you down. She tells you that since the heat came on, she has smelled smoke and when she opened the door to her apartment, she saw a haze of smoke in the hallway. She lives on the first floor. Your engine is on the scene and is positioned at a hydrant down the block.

You go into the first floor of the building and open the interior door to the cellar. Thick smoke billows up, filling the hallway. The smoke is cool and black and smells like oil burner smoke.

Scenario Questions

1. Based on what you know so far, what information would you relay back to the chief officer?

2. At this point, what are the hazards to the occupants? What actions would you take to mitigate the danger?

3. You and one firefighter go down into the basement. You locate the oil burner. The flue pipe is dislodged from the chimney. What do you do to mitigate this emergency?

4. What safety precautions would you employ while shutting the fuel supply at the oil burner?

5. What action can you take to prevent this from occurring again, possibly with more serious consequences?

6. What would you have the engine company do at this incident? Should they stretch a hoseline into the house?

7. You have shut the burner down and opened the basement windows. Can you let the occupants back into the building at this time?

Answers

1. False. We should confine our role to eliminating the hazard. Doing repair work leaves the department and the individual firefighter liable for any injury or damage that results from the repair work.
   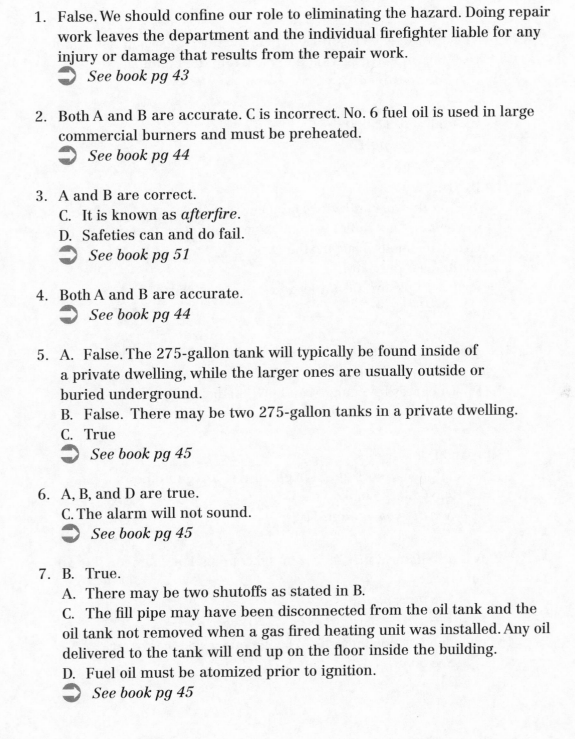 *See book pg 43*

2. Both A and B are accurate. C is incorrect. No. 6 fuel oil is used in large commercial burners and must be preheated.
   See book pg 44

3. A and B are correct.
   C. It is known as *afterfire*.
   D. Safeties can and do fail.
   See book pg 51

4. Both A and B are accurate.
   See book pg 44

5. A. False. The 275-gallon tank will typically be found inside of a private dwelling, while the larger ones are usually outside or buried underground.
   B. False. There may be two 275-gallon tanks in a private dwelling.
   C. True
   See book pg 45

6. A, B, and D are true.
   C. The alarm will not sound.
   See book pg 45

7. B. True.
   A. There may be two shutoffs as stated in B.
   C. The fill pipe may have been disconnected from the oil tank and the oil tank not removed when a gas fired heating unit was installed. Any oil delivered to the tank will end up on the floor inside the building.
   D. Fuel oil must be atomized prior to ignition.
   See book pg 45

8. True. If hot combustion chamber walls have vaporized some of this fuel, explosive re-ignition can occur.
   ⟶ *See book pg 46*

9. The draft may be insufficient to deliver flue gases up the chimney, resulting in flue gases spilling back into the building.
   ⟶ *See book pg 47*

10. C. 10,000 to 12,000.
    ⟶ *See book pg 47*

11. B. True.
    A. The thermostat controls the ignition cycle in hot air and steam systems but not in hot water systems. In hot water systems, the circulator motor controls the ignition cycle.
    C. It has a red plate.
    D. It's 15 psi for hot water and 2 psi for steam
    ⟶ *See book pg 48*

12. A and C are true.
    B. It may cause the boiler to crack.
    D. The pressure change can blow the boiler apart.
    ⟶ *See book pg 48 and 49*

13. A and D are true.
    B. Puffback occurs at the beginning of an ignition cycle.
    C. Heavy black smoke.
    ⟶ *See book pg 49 and 50*

14. It is a structural fire. If it is in the combustion chamber, it is an emergency.
    ⟶ *See book pg 51*

15. The flue has been dislodged or the burner door is open.
    ⟶ *See book pg 51*

16. All are correct.
    ⤷ *See book pg 51*

17. It may be necessary to extinguish the fire burning in the firebox.
    ⤷ *See book pg 51*

18. Pulsation.
    ⤷ *See book pg 52*

19. This bracing of the burner door might result in explosive failure of the door or burner.
    ⤷ *See book pg 52*

20. The white ghost.
    ⤷ *See book pg 53*

21. A, B, and C are true.
    D. The smoke will turn pearly white.
    ⤷ *See book pg 54*

22. C is accurate.
    A. The change in color of the smoke may indicate the onset of the white ghost. There could be a devastating explosion accompanied by a fireball at any time.
    B. It is easy to breathe, but it contains carcinogens from which you must protect yourself. SCBA should be used.
    ⤷ *See book pgs 53-57*

23. A. True.
    B. The tank can fail explosively. Other safeties won't prevent it.
    C. It is the sign of a defective temperature and pressure valve.
    D. It is the sign of a defective low water cutoff.
    ⤷ *See book pgs 61, 62, and 63*

24. A is correct.
   B. Allowing cold water into the tank could result in tank rupture.
   ➲ *See book pg 63*

25. Both A and B are true.
   ➲ *See book pgs 63 and 64*

Scenario Answers

1. Tell the chief that there is probably an oil burner emergency in the cellar. Your reason for giving this report is because the oily smelling smoke is cool, indicating that the fire is probably not yet out of the firebox.

2. There is the possibility that the emergency might escalate to a structural fire and there is a danger to the occupants from the CO-laden oil burner smoke. While checking the basement and the oil burner, get the chief's CO meter and determine the level of CO in the building. If it has reached dangerous levels, remove the occupants to safety. This may include forcing entry into apartments if the occupant does not answer the door. Remember, life is your first priority. It may be counterproductive to bring occupants out from their apartments when the hall is full of smoke. Sheltering in place may be the best choice at this time or you can close the cellar door and evacuate the occupants before reopening it. You must, however, check on all of the occupants and take CO readings as soon as practical.

3. A. Shut the emergency shutoff switch to the burner. It should be located just outside or just inside the oil burner room or on the wall of the basement stairs.
   B. Shut the fuel supply at the burner and at the fuel storage tank.
   C. Check for fire extension. Has burning fuel been blown out of the firebox? Have hot flue gases ignited any nearby combustibles?

4. Have the firefighter shutting the fuel supply stay low and away from the front of the burner, just in case there is a puffback and a flame blows out the door or peephole or the door blows off of the burner. Also, if the burner were a large commercial burner or dual-fuel burner, check for the presence of a gas line that might have been damaged. Leaking gas might be complicating this seemingly routine oil burner emergency.

5. Issue a repair order to have the burner checked and fixed by a licensed repairman before the burner is used again.

6. The smoke is cool; this is an emergency, not a fire. Your engine is correctly placed at the hydrant. The hydrant should be tested and the pumper hooked up to the hydrant. A line probably will not be needed, but the engine should stretch a precautionary line to the front door. If the smoke coming up the stairs were hot, a charged line would be needed in the basement right away.

7. Before you let residents back into the structure, check the building for dangerous levels of CO in the apartments, halls, and the basement. If the levels are high, keep the occupants out of the building while you perform additional ventilation. This ventilation may require the use of fans to vent below-grade areas.

3

Natural Gas Fires and Emergencies

Questions

1. **Which of the following statements are false?**

   A. Natural gas has a distinctive odor.

   B. Mercaptan is added to natural gas to odorize it.

   C. Natural gas is odorized to enable us to smell gas at levels as low as 10% in air.

2. **Which of the following gasses are heavier than air and will collect in low areas: natural gas, propane, or butane?**

3. **Match the flammable gas with its flammable range.**

   A. 2.1–9.5%   1. Natural Gas

   B. 4–14%   2. Propane

   C. 1.6–8.5%   3. Butane

4. Correct the following statement:

A. Low-pressure gas is often found in new sections of cities.

B. It is necessary for natural gas to fill an entire room before a combustion explosion can take place.

C. You should expect to find a gas vent pipe and a regulator on a low-pressure system.

D. In the event of regulator failure, a proper operating vent will cause all of the excess gas flow to vent to the exterior of the building and not into the building.

5. Which of the following statements about combustion explosions are true?

A. For a combustion explosion to occur, the gas must be confined, must be within its flammable limits, and must encounter an ignition source.

B. In a combustion explosion, heated air expands rapidly, tripling in volume for every 459°F of increase.

C. A combustion explosion can result in pressure buildup of 60–110 psi.

D. Most buildings will not suffer structural damage as a result of a pressure increase below 20 psi.

6. Which of the following statements are true?

A. The odorant mercaptan can be scrubbed out of natural gas if the gas is leaking up through sandy soil.

B. This can result in an odorless gas leak in a structure.

C. Experienced firefighters are well equipped to investigate reported gas odors without a combustible gas indicator (CGI).

D. Your local utility company should be notified of all verified gas leaks.

7. **Pick which statements are true.**

   A. Over-pressurization of a natural gas appliance can result in the gas flame being blown out (resulting in a gas leak) or the flame growing to dangerous heights.

   B. You are at a serious gas leak where you find gas levels in the explosive range. Since the gas meter is located inside the building, the safest action for you to take is to shut the gas off at the meter.

   C. The curb valve is usually located on the sidewalk under a square or round metal box with a square or round cover.

   D. The curb valve is shut with a special wrench, which in most cases will be turned one full turn to shut the valve.

8. **Delayed ignition occurs when natural gas builds up in the combustion chamber of an appliance and then ignites. When this occurs, what is the hazard?**

9. **A natural gas leak has ignited inside of a building. Which of the following tactics are correct?**

   A. If a gas fire is exposing combustibles, protect them with a hoseline. A solid stream works well for this purpose.

   B. Extinguish the burning combustibles if doing so won't extinguish the flames.

   C. At an outside gas fire, if the flame is impinging on a building, operate a solid stream on the exposure.

   D. When natural gas is burning, extinguish the flames with your hose stream as soon as possible.

10. **When responding to a reported gas leak, where should the apparatus *not* be positioned?**

11. It is possible for odors to enter a home from the sewer lines through an open trap or as the result of an occupant pouring an odorous liquid down the drain. What action can you take to alleviate this problem?

12. It is important to use a combustible gas detector with a __?__ readout because it will indicate whether or not the gas is in the explosive range. Fill in the blank.

13. All but one of the following statements are true. Pick the *incorrect* statement.

    A. A slight gas odor in the hall of an apartment might indicate a more serious leak in one of the apartments.

    B. Slip the gas detector's probe under the crack of the door near the floor to get an accurate measure of the gas in a locked apartment.

    C. If the door has a security peephole, remove it and insert the gas detector's probe into the room to take gas readings.

    D. If after taking readings at all of the doors in a hallway, you have not found the leak, you may have to force doors.

    E. Consider that the gas odor may be coming from an exterior source or from another floor.

14. Which of the following statements are true?

    A. Checking the individual meters in the meter room of an apartment building may indicate which apartment has a gas leak.

    B. Applying a soapy solution to a gas pipe connection that you suspect is leaking will stop the leak.

    C. Moving an appliance to check for a leak might actually cause a leak in older flexible connectors.

    D. The Consumer Products Safety Commission (CPSC) reported that 38 deaths and 63 injuries are attributable to the older uncoated brass type flexible connector.

15. **Which of the following statements are *incorrect*?**

    A.   When called to a gas leak in a structure, shutting down a single valve or meter will always solve your problem.

    B.   Even if you feel you can safely handle a gas leak in a home by shutting a meter, you should still notify the gas company of the leak and of valves shut by firefighters.

16. **Arrange the following gas shutoffs in priority order with the one that you would try to shut off first at the top of the list.**

    A.   Meter wing cock

    B.   Street valve

    C.   Appliance quarter turn shutoff

    D.   Curb valve

17. **Which of the following are true statements?**

    A.   The meter wing cock is closed by turning it one half turn (usually clockwise).

    B.   When the rectangular nut is perpendicular to the gas piping, it is in the "on" position and gas is flowing through it.

    C.   If the appliance shutoff valve won't turn, apply more pressure. They are often stuck but can be freed with enough pressure.

    D.   In private homes, the meter will most likely be located where the gas enters the building.

18. **Pick the *incorrect* statements.**

    A.   A curb valve is only found if high-pressure gas is supplied to a building.

    B.   The curb valve, if present, can usually be found on the house side of the curb.

    C.   Once you have fixed the gas leak, it is OK to turn the gas supply back on.

    D.   The street valve is located on the street side of the curb and should only be shut by the utility.

    E.   Closing the street valve might disrupt service to a substantial number of gas customers.

19. Name some ignition sources that could cause a spark capable of igniting a natural gas leak in a building.

20. In what areas of a building might gas still remain even after you ventilate the building?

21. Which of the following statements about venting natural gas are *incorrect*?

    A. Natural ventilation will usually be adequate to remove natural gas from a building and is the safest, easiest option.

    B. Positive pressure ventilation (PPV) used to push the gas out of the building is always a safer method of ventilation than natural ventilation.

    C. Cross-ventilation should speed up the ventilation process.

    D. Simply opening windows both at the top and the bottom will not in most cases vent an area.

    E. If a multistory building is full of gas, you should vent the highest level of the building first, and then as you descend, vent the lower floors.

22. Which of the following statements about outdoor gas leaks are *incorrect*?

    A. A sewage treatment plant can emit odors that might be mistaken for a natural gas odor.

    B. Natural gas that has leaked up from under the ground for a long time will cause the foliage in the area to grow rapidly and take on a bright green color.

    C. A high-pressure pipeline leak can be plugged by applying hand pressure to the leak.

23. **Name several items that can be used to temporarily stop or slow a low-pressure gas leak.**

24. **Which of the following statements is *incorrect*?**

    A. If you do plug a gas leak yourself or shut a gas valve, notify the utility personnel of actions taken.

    B. Gas supply pipe is made only of metal.

25. **Select the correct words to make each statement true.**

    A. Metal pipe <u>is non-conductive / conductive</u> and gas leaking from a metal pipe <u>does / does not</u> pose a static electricity hazard.

    B. Plastic pipe is <u>non-conductive / conductive</u> and <u>will / will not</u> allow a static charge to build up as the gas flows freely from the break.

    C. The static charge on plastic pipe can be as high as <u>600V / 1400V</u>.

    D. If a firefighter approaches or touches a leaking plastic pipe, the resulting static spark <u>can / can not</u> ignite the leaking natural gas.

26. **The hazard of static electricity discharge from a leaking plastic gas pipe is real. Utility workers take special precautions when working with plastic pipe. Which of the following statements are *incorrect* about these precautions?**

    A. A rag that is wet with plain water is placed on the pipe and in contact with the ground.

    B. Plain water is sprayed onto the pipe.

    C. Keep the pipe and cloth wet with plain water.

    D. Utility workers ground their tools before working on plastic pipe.

27. Once you park upwind from an outdoor gas leak, what action should you take to ensure that your apparatus does not present an ignition danger to the gas?

28. When natural gas is escaping from a broken pipe in the street, we still must check the nearby buildings for the presence of escaping gas. Which of the following statements about this situation are *incorrect*?

    A. Gas always percolates up and out of the ground and thus will not travel long distances underground to distant buildings.

    B. Natural gas can be trapped underground by pavement or even by a layer of frost.

    C. Gas trapped as in answer B can travel long distances before escaping into the atmosphere or into a building.

    D. Sandy soil will slow down the travel of escaped gas and prevent it from migrating from the leak.

29. Which of the following outdoor gas leak tactics are stated *incorrectly*?

    A. Identify the hazard area and set up barriers to exclude the public from the danger zone.

    B. Stretch lines into the gas cloud and operate a fog line to prevent ignition.

    C. Evacuate all in the hazard area.

    D. Determine where the leak is.

    E. Determine what resources are needed.

    F. Give an accurate description of conditions to the utility company.

Topics for Drill

1. Call your local utility company and arrange to have a representative come to the firehouse to give a drill on the hazards of natural gas. Most utility companies will be happy to provide this service to your department.

2. Identify where in your district you might encounter high-pressure gas distribution lines and high-pressure gas service. Also identify where you would find low-pressure distribution lines and low-pressure service.

3. Determine if you have curb valves in your district. Go out and locate some of these valves and explain to your firefighters how and when you would use the valve to shut the gas service to a building. Also point out street valves to them and explain why only the utility should shut them.

4. List the various flammable and combustible gases that you might encounter in your district.
   A. What are the hazards and properties of each gas?
   B. Where would each gas be found?
   C. What safety precautions must you take when a fire or emergency involves or threatens to involve each gas?

Scenario: Natural Gas Emergency

You are a lieutenant of a ladder truck staffed with four firefighters in addition to you. Your engine company has a lieutenant and three firefighters. You have been dispatched to the report of a leaking gas main at a restaurant along with one engine and a chief. The chief carries a CGI. Your ladder truck arrives first on the scene and you hear the loud sound of escaping gas and smell a heavy odor of gas in the air. A backhoe has pulled up and torn a high-pressure gas service line midway between the street and the restaurant it serves. The restaurant is in a 1-story strip mall.

Scenario Questions

1. What information should you relay to incoming units over your apparatus mobile radio?

2. What additional help if any would you call for?
    *   Fire department

    *   Other

3. How large an area would you consider to be in the danger area?
    - What would you include in the danger area?

4. Describe how and why the escaping gas could migrate into nearby buildings.
    - Aboveground

    - Underground

5. You determine that the gas service line is plastic. One of your firefighters says he can crimp the gas line and stop the flow of the gas. Is this a good idea? Why/why not?

6. The truck lieutenant suggests that one of his firefighters close the gas valve in the street. Is this advisable? Why/why not?

7. You have your firefighters checking the nearby buildings. One firefighter reports to you that the gas has not entered the building he is checking. He does not have a CGI. He says that there is no gas odor at all in the building and that it is safe. Should you accept this report? Why/why not?

Answers

1. A. False. An odorant must be added to natural gas so that a leak can be detected by the user.
    B. True.
    C. False. We can smell odorized gas at levels as low as $1/10^{th}$ of 1%.
    ⮕ *See book pg 67*

2. Propane and butane are heavier than air and will collect in low areas. Natural gas is lighter than air and will rise.
    ⮕ *See book pg 68*

3   A. 2. Propane.
    B. 1. Natural gas (sometimes described as 5–15%).
    C. 3. Butane.
    ⮕ *See book pg 68*

4. A. Found in old secions.
    B. Natural gas can collect in only a portion of the enclosure and still result in a combustion explosion.
    C. High-pressure system.
    D. Some of the excess gas will flow into the building's piping system and into the appliances.
    ⮕ *See book pg 69-74*

5. A and C are true.
    B. It doubles in volume, not triples.
    D. Most will suffer structural damage if the pressure rises as little as 1 psi.
    ⮕ *See book pg 71*

6. A, B, and D are true.
    C. False. Even an experienced firefighter will not be able to smell natural gas once the odorant has been scrubbed out of it. You need a CGI.
    ⮕ *See book pg 72*

7. A and C are true.
    B. Shut the gas at the curb valve.
    D. The curb valve is shut by one-quarter turn.
    ➲ *See book pg 74*

8. When the gas ignites, a small explosion occurs, projecting a flame beyond the combustion chamber, possibly igniting nearby combustibles.
    ➲ *See book pg 74*

9. A. False. A narrow fog stream should be used.
    B. Correct.
    C. False. Use a fog stream.
    D. False. Extinguish the fire by shutting the gas.
    ➲ *See book pg 77*

10. It should not be positioned directly in front of the building. Position it out of harm's way, out of the potential collapse zone.
    ➲ *See book pg 78*

11. If possible, seal the trap and then flush the building's sewer piping by flushing the toilet and running water into the sink. Sealing the trap prevents odors from entering the building, and flushing the pipes will push an odorous liquid out of the building.
    ➲ *See book pg 78*

12. Digital.
    ➲ *See book pg 79*

13. A, C, D, and E are all true.
    B is false. Natural gas rises and measuring at the floor level is not a good way to get accurate readings.
    ➲ *See book pg 79*

14. A, C, and D are true.
    B is false. If it bubbles, it indicates the presence of a leak.
    ➲ *See book pgs 80 and 81*

15. A. False. There may be more than one source of leaking gas in the building.

    B. True. They will have to test for other leaks and ensure the system is safe before it is turned back on.

    ⟳ *See book pg 81*

16. 1. C

    2. A

    3. D

    4. B (only the gas company should shut this valve.)

    ⟳ *See book pg 81*

17. A. False. Turn the wingcock one-quarter turn.

    B. False. When the nut is parallel, the gas is flowing.

    C. False. If the valve won't turn, go to a more remote shutoff (usually the meter).

    D. True.

    ⟳ *See book pg 82*

18. B, D, and E are true.

    A is false. It may also be found in some low-pressure systems.

    C is false. Only the utility company should restore a building's gas service once we shut it off. (Pilot lights must be relit and the piping tested.)

    ⟳ *See book pg 82*

19. Here are some. There are more.

    • Your hand light and radio unless they are intrinsically safe. (Turning your thermal imaging camera on or off can also supply the ignition source.)

    • Electrical switches.

    • Static sparks created as you walk over a carpeted floor and then touch a metal object.

    • Pilot lights.

    • Open flames.

    • Heating unit. (Thermostat can kick in when you open the entrance door and a cool breeze hits it.)

    • Pulling the electric meter. (Gas can travel up the electrical conduit into the meter. A spark can be created when you pull the meter.)

    • Ringing telephone.

    ⟳ *See book pg 85*

20. Closets, cabinets, attics, walls and other enclosed spaces that might trap gas and prevent it from venting from the building.

     ➲ *See book pg 86*

21. A, C, and E are true.
     B is false. You may be making the situation more hazardous by pushing the gas into other areas of the building.
     D is false. This will in most cases vent the area.

     ➲ *See book pg 86*

22. A. True
     B. False. The foliage will eventually turn brown and die.
     C. False. A low-pressure leak may be stopped this way, but not a high-pressure leak.

     ➲ *See book pg 87*

23. Duct sealant, duct tape, wad of paper, or rags. If the pipe has been completely severed, a wooden plug.

     ➲ *See book pg 87*

24. A is true.
     B is false. Metal or plastic.

     ➲ *See book pg 87*

25. A. Conductive, does not.
     B. Non-conductive, will.
     C. 1400V.
     D. Can.

     ➲ *See book pg 87*

26. A, B, and C. Soapy water is used. Soapy water breaks the plastic's surface tension whereas plain water will simply bead up, making it insufficient path to ground.
     D is true.

     ➲ *See book pg 88*

27. Shut down your apparatus and electrical system if it is not needed for your operation. Consider that though you are parked in a safe area now, the wind may shift, exposing your apparatus to the gas cloud.

    See book pg 88

28. B and C are correct.
    A is false. Trapped gas can travel long distances underground and can enter buildings.
    D is false. Sandy soil is more porous than packed dirt and allows the leaking gas to travel long distances through it.

    See book pg 89

29. A, C, D, E, and F are true.
    B is false. Entering the gas cloud even with a fog line is dangerous. Stretch hoselines to a safe area. Consider utilizing a fog stream to dissipate the gas if life or property is at risk.

    See book pg 90

Scenario Answers

1. Inform the units that you have a confirmed leak in a high-pressure gas service, to approach from upwind and to stay out of the gas cloud. Incoming units may not be able to determine wind direction and gas vapor clouds are not visible. Recommendation: Direct the incoming units to a location that you have determined to be safe.

2. Do you have enough personnel on the scene to check the strip mall and adjoining structures for the presence of gas and to evacuate the occupants if necessary? Will your requested response bring enough combustible gas indicators (CGIs) to the scene? If not, ensure that units with CGIs respond. Make use of the utility personnel with their CGIs. Once you set up a danger area and evacuate people from it, you will need the police to maintain this area clear of civilians and traffic. You will need a fire officer dedicated to keeping firefighters out of the danger area.

   Request the immediate response of the gas company, and relay the exact location of and the nature of the leak. This will enable the utility control room personnel to begin identifying the quickest way to shut down and make safe the gas line. They will either have to identify, find, and close the appropriate valves to stop the flow of gas, cut and cap the line, or squeeze-off the line.

3. When the backhoe pulled up the gas line, not only was the pipe ruptured at point of impact, one or both ends of the service line may have possibly been damaged. That means that either the pipe was pulled away from the strip mall or pulled out of the distribution main in the street or both. Gas can be leaking out of the pipe near the building wall and migrating into the building. It is also possible that gas is leaking in the street if the service line was pulled out of the main. This leaking gas can migrate under the cement or asphalt or even under a layer of frost, eventually entering a building some distance away. This is in addition to the gas leaking at the obvious break.

The danger area could include more than just the immediate area around the damaged line. You will have to check for the presence of gas in buildings, and utility personnel will have to check for gas migration underground.

4. Escaping gas could migrate into nearby buildings in a number of different ways.

Aboveground

- The gas escaping into the air could be drawn into the building by its ventilation system.
- Natural air currents could allow gas to enter open windows or doors.

Underground

- It could seep from the dislodged pipe into the strip mall and from there spread to adjoining occupancies of the mall. Is there a common cockloft or an air conditioning or heating system that could channel the gas to other stores?
- It could travel underground and enter one of the various stores into the strip mall.
- It could enter the sewer system and travel great distances to distant parts of the mall or to unattached buildings.
- Leaking gas may migrate into the underground electrical system (manholes, transformers, etc) or sewers and could then follow the conduits into structures.

5. It is not a good idea. Do not let firefighters touch or go near a ruptured plastic gas line. Gas flowing out of a plastic gas line creates a static electricity charge on the gas pipe. Touching the pipe or even approaching it could cause a static discharge sufficient to ignite the escaping gas. Wait for the gas utility to shut off the gas.

6. The curb valve, if present, will control the flow of gas into the building, and if shut, it will cause no harm. A street valve controls the flow of gas through a main and shutting it may disrupt the flow of gas to many users in the area. It will require that the gas utility test pipes and light pilots of all of the gas users served before turning the gas back on. This will take a substantial amount of time and result in a prolonged gas outage. In addition, it is possible that the service is fed in two directions and closing the street valve will not affect the desired shutdown. Do not shut down street valves.

7. Do not accept this report as conclusive. The odor can be scrubbed from natural gas as it passes through the soil. Additionally, prolonged exposure to a gas leak may desensitize your nose to the odor. The result could be natural gas that he might not be able to smell. You can't trust this firefighter's sense of smell as a positive indicator of the absence of natural gas. Get a CGI into the building to verify your firefighter's educated nose.

4

Water Leaks

Questions

1. When we respond to a building for a water leak, which of the following tasks should we attempt to accomplish?

   A. Locate the source of the water.

   B. Stop the flow (if possible).

   C. Remove any hazard created by the water.

   D. Repair the leak.

2. True or false? The usual cause of a flooded roof is a clogged roof drain.

3. True or false? A parapet running entirely around a flat roof will have impact on the amount of water contained on a flooded roof. In the event of a clogged drain, the presence of a parapet can confine the water, creating a swimming pool on the roof.

4. **Which of the following statements about flooded roofs are *incorrect*?**

   A. Small, flat roofs usually slope down from front to rear, whereas large flat roofs slope toward drains that are typically located at the low points on the roof.

   B. If there is only a small flood on the roof, the drain will usually be within the flooded area.

   C. If the entire roof is flooded, it will be easy to find the drain.

   D. Drainpipes always run down the side of the building and are easily spotted from the street.

   E. If you can't find the drain, contact the building maintenance personnel. They should be able to help.

5. **A flat roof that is 20 ft by 40 ft with a parapet and a clogged drain will contain __?__ cubic feet of water if the water level is 6 in. deep. Fill in the blank.**

6. **How much does the water on the roof in question 5 weigh?**

7. **Which of the following statements are *incorrect*?**

   A. If the roof is in danger of collapse due to a buildup of water on the roof, you should be thinking of evacuation and a means to drain the roof.

   B. Jumping onto a flooded roof from a ladder will create a dangerous impact load on the already overloaded roof.

   C. If you decide to operate on such an overloaded roof, use as many firefighters as possible to get the job done quickly.

   D. When you find a clogged drain covered by water, reach into the drain and try to clear the drain with your hand.

8. **Describe three methods mentioned in the book for draining water from a flooded roof.**

9. **If water is leaking from an apartment on the 3rd floor down into the apartment on the 2nd floor, you may have to get into the 3rd floor apartment to stop the leak. Which of the following statements accurately describe the hazards and/or benefits of forcing entry into that apartment?**

    A. The occupant may be asleep or drugged. If he awakes, he may try and defend himself. If he has a weapon, the result could be deadly to the firefighters.

    B. The occupant could be unconscious and in need of medical assistance. Not forcing entry could be deadly to the occupant.

10. **Which of the following statements are correct?**

    A. If water is leaking from the ceiling in an apartment, then the source of the leak is on the roof.

    B. If you see no obvious source of the leak mentioned in question A in the upstairs apartment, but you hear water flowing, the leak might be under the sink or coming from a broken pipe in a wall.

11. **Once you are in the apartment with the leak, you should try and stop the leak. Which of the following statements is *incorrect*?**

    A. The most likely source will be an overflowing sink, tub, or toilet.

    B. If the problem is the result of a clogged sink, removing the clog should solve the problem. Carrying a plumber's snake with you into the apartment will make this task easier.

    C. If the toilet tank is overflowing, shaking the handle may solve the problem.

    D. If shaking the handle does not solve the problem, remove the tank top, and bend the rod so that the float sits lower in the tank.

    E. If the problem is a stuck flushometer, tighten the hexagonal fitting if loose. If that does not help, try tapping it with the back of an ax.

12. To stop the flow of water from a toilet or sink, you can close the shutoff valve located under the sink or toilet. If the leak is from a broken pipe, this tactic may not work. Which of the following methods of stopping the flow from a broken pipe are *incorrect?*

    A. Wedge a small wooden plug into the opening.

    B. A pencil or even a golf tee may be jammed into the pipe.

    C. Wrapping the plug with cloth or tape before inserting it may help tighten the fit.

    D. Wrap duct tape around the break in the pipe.

13. If you can't stop the flow of water by any other method, you should go to the cellar and shut the control valve that supplies water to that apartment or line of apartments. In large buildings, there may be multiple shutoff valves. How can you locate the appropriate one?

14. Which of the following statements are *incorrect?*

    A. If you can't shut down the water to the apartment or line of apartments, you may have to shut down the house main, stopping the flow of water to the entire building.

    B. Shutting the house main will immediately stop the flow of water from the leak.

    C. Shutting the water supply will slow the leak and eventually stop it.

    D. Position a radio-equipped firefighter at the leak to inform the firefighter at the valve when the water flow slows down.

15. In addition to stopping the flow of water from a broken pipe, you may have to disconnect the electricity to the apartment or to the entire building. True or false?

16. **Which of the following statements about dewatering operations are *incorrect*?**

    A.   An eductor is a dewatering device that uses Bernoulli's principle and atmospheric pressure to siphon water.

    B.   An eductor requires a minimum of manpower to operate and can be left with a single firefighter to operate while the pumper and remaining crews respond to other emergencies.

    C.   An eductor can safely pump water that contains small particles of debris

    D.   During a storm, you may not want to tie up your department's resources removing water from a flooded basement. You can restrict yourself to mitigating hazards by removing electricity, shutting down gas supply, containing any resultant fuel spill and evacuation of occupants.

    E.   After a storm when your response load is less, you can return to assist the homeowner in removing the accumulated water from his basement.

17. **Pick the false statements.**

    A.   Several inches of water on the cellar floor can become deadly to firefighters if there are pits or other levels hidden below the surface of the water.

    B.   Some oil burners are installed in pits, and these pits can fill with water.

    C.   As you move forward through a flooded basement, shine your light in front of you to look for submerged pits.

    D.   A small child can drown in a few inches of water.

18. **The following is a list of problems or hazards that can occur as a result of a flooded basement. Which of these statements are true?**

    A.   If the water level reaches the level of the gas burner, the pilot light can be extinguished.

    B.   If water reaches the electric outlet, responding firefighters could be in danger of electrocution.

    C.   If the water level rises to the level of the oil tank, water can seep into the tank and contaminate the oil.

19. **Which of the following statements are *incorrect*?**

    A. In vacant buildings, piping is often removed for *mongo* or salvage.

    B. If the source of the water is a burst pipe, you must wait for the water department to shut down the water supply to the house.

    C. If the leak is from flexible metal tubing, you can crimp the tube.

    D. When shutting down a water valve, in most cases the wheel or lever is turned clockwise.

20. **To remove water from the floor of a flooded basement, breaking the toilet bowl at the floor will quickly get rid of the water. Is this a good idea? Why/why not?**

21. **One way to remove water from a flooded basement is to find and remove the ___?___ from the waste pipe. Fill in the blank.**

22. **After you remove the toilet or the cleanout plug to remove water from a flooded basement, how can you prevent the opening from clogging with floating debris?**

23. **Even after we stop the flow of water and remove the accumulated water from the building, the occupants may still be in danger. Which of the following statements is *not* true?**

    A. Water that is still in the ceiling, in walls, and on wires and appliances can cause switches, appliances, and even countertops and floors to become electrically charged.

    B. The hazard noted in answer A can occur after you leave if you don't take appropriate precautions.

    C. You must ensure that the power to the entire building has been shut down.

    D. The occupant must understand that he is not to restore the power until the water has dried, and the system has been checked by a licensed electrician.

24. Upon discovering a flooded basement, assign a firefighter to enter the basement to open the main breaker or pull the main fuse. True or false?

25. Which of the following statements are *incorrect*?

    A. Weakened, water-soaked ceiling plaster, sheetrock, or dropped ceiling can fall at any time, possibly causing serious injury.

    B. Poke a sagging ceiling with a pike pole to relieve the buildup of water above it.

    C. Always remove as small a portion of the ceiling as possible to save the homeowner from the expense of replacing it.

    D. Move or cover furniture before relieving the ceiling of water.

26. Pick the false statement.

    A. Water leaking down an elevator shaft can cause the elevator to malfunction.

    B. If water has entered the shaft, check the elevator for trapped occupants.

    C. Water can affect the car's safety circuits and can result in sudden, unintended movement of the elevator car.

    D. If water has entered a multi-car elevator shaft, and people are trapped in one of the cars as a result of a malfunction, use an adjoining car to remove them.

27. A burst water main can create several hazards. Which of the following is *incorrect?*

    A. If not repaired, a water-main leak can wash away the soil underneath the concrete and asphalt.

    B. The result could be the eventual collapse of the sidewalk and street.

    C. If a gas pipe runs through the area, it too could be undermined and could break.

    D. The only utility that must be notified of a water main break is the water company.

28. **Listed here are some hazards associated with a burst water main. Pick those stated *incorrectly*.**

    A. The possibility always exists that the pavement has been undermined and that the weight of a firefighter, your apparatus, or other vehicles will collapse it.

    B. It is also possible that the ground has already collapsed and that the collapsed section isn't visible due to the water that has filled up the void, leaving what looks like a puddle.

    C. The presence of a sewer manhole in the flood area presents no additional hazards. In fact, the water may drain into the manhole, lessening the flooding problem.

    D. An undermined gas main might break, resulting in a gas leak.

Topics for Drill

1. Locate a flat-roofed commercial or residential building in your area and point out the location of roof drains to your firefighters and explain how you would dewater the roof if the need arose. Remember to point out the hazards involved with using your hand to clean out a clogged drain.

2. Go into the cellar of a large commercial or residential building in your area and point out the various hazards (to firefighters and occupants) that might be encountered if the basement flooded. Include the following if appropriate:

   A. Pits

   B. Sub cellar

   C. Fuel oil tanks

   D. Gas appliances and meters

   E. Electrical boxes and outlets

   F. Elevators

   G. Descending stairs

3. Compile a list of agencies and private companies that you might call upon for assistance in the event of a large water main break on a densely populated street. Assume that businesses, multiple dwellings, and private homes are flooded.

   A. What hazards would exist in the street?

   B. Describe what tasks firefighters might be called upon to do in such an emergency.

   C. How can the various agencies assist you?

   D. How might the various utilities be affected?

Scenario: Water Leak Emergency

You are a chief. Your department staffs its engines with an officer and three firefighters and its trucks with an officer and four firefighters. It has been a cold winter and today is no exception. The mercury is stuck at 5°F. You are out of quarters when you receive a call for a flooding condition in your downtown area. Your downtown area is the older part of your city.

You are close to the reported location when you are notified, so you arrive first. Your response is an engine and a truck that will only take a few minutes to arrive.

What you see on arrival is a 2-in. deep river of water flowing across your main street and into several older buildings. A parked car is half-sunken below the water level with its front sticking up into the air at a 60° angle.

Scenario Questions

1. What life hazards to civilians and firefighters exist at this incident and what action must be taken to protect all?

2. Is your response adequate?

3. Your gas utility responds and tells you that a 4-in. cast iron gas main is in the vicinity of the water leak. What problems does this pose?

Answers

1. A, B, and C are accurate.
   D is not accurate. We are not repair men. We are there to mitigate the hazard.
   ⟳ *See book pg 94*

2. True.
   ⟳ *See book pg 94*

3. True.
   ⟳ *See book pg 94*

4. C. If the entire roof is flooded, it will be more difficult to find the drain.
   D. Drainpipes often run into the building and not down the side. They can be difficult to locate.
   ⟳ *See book pg 94*

5. 400
   Length × width × height = cubic feet of water
   20 ft × 40 ft × .5 feet = 400 cu ft of water.
   ⟳ *See book pg 95*

6. 24,960 lbs.
   Cubic feet of water × 62.4 lbs (weight per cubic foot) = weight of water
   400 × 62.4 = 24,960 lbs.
   ⟳ *See book pg 95*

7. C. Allow as few firefighters as possible on the roof.
   D. Never attempt to free a clogged drain by hand. The suction created when the drain is cleared can pull your hand down into the drain.
   ⟳ *See book pg 95*

8. Here are three methods:
   - Submerge a length of rubber hose and let it fill with water. Kink one end. Let that end hang over the edge of the roof, lower than the roof. Remove the kink, and water will siphon off of the roof through the hose.
   - Stretch a rubber hose up the side of the building and onto the roof. Charge it, then shut down the pump. Disconnect the hose from the pumper. The water flowing back down the hose will create suction and drain water from the roof.
   - Use a salvage pump.

   ⮑ *See book pg 95*

9. Both A and B are true statements.

   ⮑ *See book pg 97*

10. A. This is true for a top-floor apartment. On lower floors, the leak is probably coming from the apartment above.

    B. True

    ⮑ *See book pg 97*

11. B. Don't start snaking the pipe to clear the clog. You are not responsible for repairs.

    ⮑ *See book pg 97*

12. D. The plug can be held in place by duct tape.

    ⮑ *See book pg 97*

13. To locate the appropriate valve, listen for the sound of running water and feel the pipes. The one with water running through it will be colder than the rest. (That is, if it is a cold water pipe that is leaking.)

    ⮑ *See book pg 97*

14. B. It may take time for the water to drain if the leak is on the lower floor of a multistory building.

    ⮑ *See book pg 97*

15. True.
    ⮕ *See book pg 98*

16. A. The eductor uses the venturi principal and atmospheric pressure.
    B. The eductor requires a pumper to supply it with water.
    ⮕ *See book pg 98*

17. C. As you move forward through a flooded basement, probe the area in front of you with a pike pole or other tool before each step.
    ⮕ *See book pg 99*

18. A and B are true.
    C. The fuel oil tank can begin to float, causing breaks in the piping and resulting in oil spills.
    ⮕ *See book pg 99*

19. B. You may be able to stem the flow from a burst pipe by plugging it or shutting down a valve.
    ⮕ *See book pg 100*

20. This is OK to do at a fire to reduce the water load on an overloaded floor. It is an emergency tactic and should not be done in non-emergency situations. It is better to remove the bowl intact with the same result. That way, the homeowner will not have the added expense of replacing it.
    ⮕ *See book pg 100*

21. Cleanout plug.
    ⮕ *See book pg 100*

22. Place some sort of screen over the hole to prevent floating debris from clogging the opening.
    ⮕ *See book pg 100*

23. C. Ensure the power to the all the *affected areas* has been shut down. This may in fact include the entire building but not necessarily.
    ➩ *See book pg 100*

24. False. It isn't prudent to allow a member to stand waist deep in water when tripping circuit breakers or pulling fuses. Have a utility worker pull the meter or cut the wire from the exterior to remove the power.
    ➩ *See book pg 100*

25. C. Once the excess water has been relieved, it may be necessary to remove a large portion of the ceiling to ensure that it won't fall down at a later time.
    ➩ *See book pg 101*

26. D. Firefighters should not use an elevator once water has entered its shaft.
    ➩ *See book pg 101*

27. D. Also notify the gas utility of a water main break.
    ➩ *See book pg 102*

28. C. The escaping water may dislodge the manhole, and an unsuspecting firefighter cold step into the unseen hole.
    ➩ *See book pg 104*

Scenario Answers

1. Life hazards to civilians and actions to be taken:

   A. A search of the involved area must be conducted.

   B. The pavement has collapsed under the car that is sinking into the street. This car must be searched quickly, before it disappears. Be careful that firefighters do not sink into the hole with the car.

   C. How much of the roadway and pavement is compromised? Both civilian and fire vehicles could become entrapped by collapsing roadway. Traffic must be stopped, the area taped off, and responding FD units directed to a safe area.

   D. Is water entering buildings? Do these buildings have cellars or basements? If so, is there anyone in these areas? Searches will have to be made quickly to rescue any endangered occupants.

   E. Has the escaping water undermined buildings, walls, light and electrical poles, or any other structures? A rapid assessment of this hazard must be made and evacuation of endangered areas initiated, if necessary. Call building department engineers to the scene to make a stability determination if necessary.

   F. Is the undermining of the roadway threatening underground utilities? Is a gas line in danger of breaking and are underground electric lines being flooded? Have your local utility respond to determine if there is need for concern.

   G. Will there be a problem with the building systems as water pours into surrounding properties? The electrical systems of buildings may need to be de-energized. Elevators may be stopped and, if so, will have to be searched for trapped occupants. Gas may have to be shut off and testing for leaking gas inside structures would be prudent. Rising water in the cellar or basement of structures might float fuel oil tanks. This could result in oil spills and damage to gas pipes as the fuel oil tank moves off of its moorings. It may be necessary to contain/reclaim the spilled oil.

   H. The 2 in. of water might be concealing collapsed pavement or a dislodged manhole. An unsuspecting civilian or firefighter could step into what he thinks is 2 in. of water and wind up sinking into water that is over his head. Civilians must be kept out of the flooded area, and firefighters must test each step with a tool before moving forward. Firefighters must be cautious of pits, and other openings in cellars that could fill up with water. A sinkhole or a pit might just look like a puddle to an unsuspecting firefighter. Stepping into such a flooded space could result in firefighters drowning. Firefighters should use their tools to check for solid ground in front of them as they walk into the water.

I. The cold weather will result in quick freezing of water so a slipping hazard will exist at the edges of the running water. Both civilians and firefighters could be injured in falls. Once traffic resumes, accidents may result from the slippery condition. It may be necessary to put salt or sand down after the water flow has been stopped.

J. Be alert to the possibility that the water in structures may have contacted electrical circuits and be carrying an electrical charge. Firefighters searching for victims would be in danger of electrocution. FD units can shut the gas and electric if need be, but do not let firefighters place themselves in danger of electric shock by standing in water when operating electrical switches. If necessary, have the utility company cut the power to the flooded buildings from the exterior.

2. Clearly, one engine and one truck with a total of seven firefighters and two officers will not suffice to mitigate this emergency. You will need much more.

A. You will need the water department to stop the flow of water. This may take some time. The appropriate valves must be located and shut.

B. The police department (PD) should be contacted for crowd and traffic control and, if needed, to help with search and evacuation of threatened structures.

C. Until PD is on the scene, you will have to stop pedestrian and vehicular traffic yourselves.

D. Gas, electric, and telephone service may be disrupted. Gas and electric service may have to be shut off to endangered buildings. An adequate utility response is essential.

E. You will need more firefighters to safely search the cars, the street, and the buildings and to perform other related tasks.

3. A. Gas mains, especially cast iron ones, can crack as a result of their own weight when undermined by escaping water. This would result in a major gas leak that could ignite or travel both aboveground and underground, entering buildings.

   B. The utility might need to shut the gas flow to underground pipes as a precaution before they crack.

   C. You should be testing the buildings and area for gas with your CGIs. Utilize utility personnel to assist you in this.

   D. If the pipe cracks and the gas ignites, you will have a gas fire to contend with. Cars and buildings will become possible exposure problems. Do you have the resources available should this occur, and are they positioned to be able to set up defensive hose streams to protect exposures?

   E. The broken water pipe might deny you the use of your hydrant system as a source of water. Alternate water supply would be needed to protect exposures.

5

Vehicle Fires

Questions

1. Today's shock-absorbing bumpers can be compressed by impact and may stick in the compressed position as a result of a collision. This is a hazard to responding firefighters. Which of the following statements about this loaded-bumpers condition is stated correctly?

   A. The two compressed cylinders can suddenly release and cause the bumper to fly off of the vehicle, striking nearby firefighters.

   B. One of the compressed cylinders can suddenly release and cause the bumper to swing out, still hinged on the fixed cylinder side.

   C. At one incident, a loaded hydraulic cylinder was launched from the vehicle and found embedded in the ground more than 40 ft away.

   D. Be aware that the cylinder can also explode if it is sufficiently heated.

   E. To avoid these hazards, approach the vehicle from the front or rear.

2. **Which of the following statements about automotive battery hazards are *incorrect*?**

   A.   If the case of the battery is compromised, hydrogen gas can escape from the battery.

   B.   If this gas cloud finds an ignition source, it can flash back to the battery and cause an explosion.

   C.   Hydrogen gas can develop explosive pressure above 100 psi.

   D.   The explosive range of hydrogen is 20–71%.

   E.   The simple act of removing the battery cable can create a spark that can trigger an explosion.

   F.   Hydrogen and nitrogen gases are formed as the battery charges.

   G.   Automotive batteries of today are composed of a plastic shell enclosing lead cells submerged in an electrolyte solution of nitric acid and water.

3. **Which of the following statements are true?**

   A.   To prevent sparks that might trigger an explosion, cut the positive battery cable and be sure that it doesn't touch the negative terminal.

   B.   Older flat-fixing products contained petroleum distillates that were highly explosive. The newer ones are now required to be nonflammable, so this hazard is no longer a concern at car fires.

   C.   Methyl glycol (antifreeze) is toxic if ingested but poses no hazard at car fires or accidents.

4. **Which of the following statements about car fires are *incorrect*?**

   A.   A ball of fire 50 ft long, 20 ft wide, and 20 ft high has been reported as a result of an exploding automobile gas tank.

   B.   An automobile gas tank will typically fail at the seam.

   C.   Plastic tanks, used in newer cars, may melt and spill their contents onto the ground.

   D.   Fuel-injection supply lines retain their pressure when the engine is shut down. A break in the supply line could release a cloud of atomized gasoline on anyone in the vicinity.

5. **Hydraulic fluids can pose a danger to firefighters. Which of the following statements are true?**

   A. Power-steering fluid but not automatic transmission fluid is flammable when released under pressure in the form of a spray.

   B. Their flash point is in the range of 410–460°F.

   C. Power-steering fluid in automobiles can be pressurized to as much as 200 psi.

   D. Both fluids are flammable and also caustic.

   E. The smoke generated by burning brake fluid is harmful to lung tissue.

6. **You will know if a vehicle uses an alternative fuel such as compressed natural gas (CNG), propane, or electricity by the mandatory placarding. True or false?**

7. **Which of the following statements about alternative fuel vehicle fires are *incorrect*?**

   A. Some vehicles have dual fuel capability.

   B. Fire impinging on a propane cylinder may lead to a boiling-liquid expanding-vapor explosion (BLEVE).

   C. If a propane cylinder is being heated by an impinging fire, and if the fire cannot be extinguished and the tank adequately cooled, the safest tactic may be to attack from the sides of the vehicle.

   D. Base your tactical decisions on how long the fire has been burning.

8. **Which of the following statements about CNG vehicles is *incorrect*?**

   A. If such a vehicle is on fire, shut the gas.

   B. Beware of the vent line. If gas vents and is ignited, a firefighter nearby could be engulfed in flames.

   C. A CNG tank will ignite if shot by a bullet from a high-powered rifle.

   D. CNG-powered vehicles do not present any unusual hazards to firefighters.

   E. A CNG tank exposed to fire might lead to a BLEVE if the tank is not cooled down quickly.

9. **Mark the following statements as true or false.**

   A. After forcing open a car's trunk, you should consider bending the striker plate of the trunk lock before you leave the scene of a car fire to prevent a child from climbing into the trunk and becoming trapped when the lid closes and locks.

   B. When making the initial attack on a burning car, quickly get as close as possible before opening your hoseline.

   C. If a car fire occurs on a busy street or highway, consider placing your apparatus to block the oncoming traffic, thereby creating a safe working area for fire fighters.

10. **Which of the following statements is accurate?**

   A. If the backseat of a car has been burned through or removed, you should be able to gain access to the trunk from the passenger compartment.

   B. If you suspect extension from a passenger compartment fire into the trunk, quick access for extinguishment can be gained by breaking a taillight and inserting the nozzle into the hole.

   C. It is a good idea to knock down a fire in the engine compartment before opening the hood because you can expect a sudden flare up of flammable vapors to occur as you open it.

11. **List three ways that you can gain quick access to the engine compartment to allow a hose stream to knock down an engine compartment fire.**

12. **Which of the following statements about the hood release cable are stated _incorrectly_?**

   A. Once the fire in the passenger compartment has been extinguished, try to pull the hood release cable in the passenger compartment.

   B. Answer A because this is the easiest way to open the hood.

   C. Answer A and B because the mechanism often survives the fire.

   D. If it does not work, break the grille, reach through and pull the release cable.

13. Once you have opened the hood, you must prop it open so it does not fall on and injure a firefighter. Which of the following methods of keeping the hood open is accurately described here?

    A. It can be propped up with a pike pole.

    B. You can insert the adze end of a halligan tool on the hinge and bend it about 45°. The deformed hinge will keep the hood propped open.

14. A car fire inside a garage is a structural fire, and all of the rules of structural firefighting apply. The first line should always be stretched into the garage via the front overhead door to extinguish the fire. True or false?

15. Once you place a protective line at the interior door to the garage, you must decide how you want to attack the fire. You can attack it from the interior door with the protective line, or you can attack it from the exterior roll down door with a second line. Which is generally safer?

16. The hoseline that is protecting the interior garage door must have enough slack to cover what area?

17. Which of the following statements about garage fire hazards is stated *incorrectly*?

    A. The sudden involvement of a fuel tank in this confined area can injure firefighters, cause structural damage to the garage, and promote rapid fire extension to the residence itself.

    B. The cables and springs that raise an overhead garage door can suddenly fail when they are exposed to high heat.

    C. Firefighters shouldn't enter the garage before the line is charged.

    D. If the springs fail and the door crashed down, the hoseline can burst if charged. If the line has not been charged, the closed door will prevent water from reaching the nozzle. Firefighters will be trapped inside with the burning car without water.

    E. If the tracks are warped by intense heat, the entire garage door might fall down from its tracks onto personnel below.

18. **Which of the following statements are true?**

    A. At a car fire in a garage, wedge the tip of a pike pole into the track of the overhead door to prevent it from sliding down if springs fail.

    B. You can also clamp a vise grip onto the track under the raised door to prevent it from sliding down.

    C. The precautions mentioned in A and B should be taken to prevent the overhead door from closing if the tracks fail.

    D. Treat all garage fires as hazmat incidents. Move slowly, assess the situation, and determine what is burning before putting firefighters at risk.

19. **Mark the following statements about garage fires as true or false.**

    A. If you encounter stubborn flames, it may be because some exotic fuel stored in the garage is burning.

    B. White flames that intensify when water is applied could indicate a fire involving natural gas.

    C. If you are attacking a garage fire from the interior door common to the house and garage, opening the main garage door will provide ample ventilation.

    D. Always vent the garage roof in the early stages of your attack.

20. **Which of the following statements are true?**

    A. When responding to a car fire on a highway or busy street, you must take action to protect yourself as well as the unsuspecting drivers that you will encounter. This type of response should be a single unit response. More units on the scene would add to, rather than reduce the danger.

    B. The FD always has the authority to decide to close down a highway when they are operating at a car fire on that highway.

    C. Chocking the wheels of the burning vehicle only need be done when it is on a hill.

21. A fluid spill from the vehicle will make the roadway slippery. In addition, water on the roadway in the winter can freeze and cause a dangerous slippery condition. What action can you take to remove a slippery hazardous condition on the roadway as the result of a car fire or accident?

Topics for Drill

1. Identify housing *types* in your area and determine if there is interior access to the garage from the residence. Discuss how you would attack car fires in the garage for various types of buildings. Take photos of each type and train your firefighters to recognize the different types. A good time to determine if there is interior access is when you respond on medical or other emergencies.

2. What is your policy for car fires and accidents on roadways? Are your firefighters adequately protected when working at these hazardous incidents? Review safety precautions to be taken when operating on roadways. These precautions may be different on local roads and highways, depending on the traffic and geographic layout of the site. Remember to consider the additional hazards on hills and blind curves.

3. Go to a local junkyard and get permission to drill on a few junkers. Practice forcing the trunk, forcing the hood, piercing the headlight and taillight for stream access, and removing the rear seat. While you are at it, break out your Hurst tool and practice extrication.

4. Contact your local PD and ensure that both the local police and your firefighters are aware of their authority and responsibilities at incidents on the highways and streets. Also contact other agencies that might be helpful to you at car fires on roadways. For example, sanitation can provide sand and cleanup of debris. State police are responsible for state roads.

Scenario: Vehicle Fire Emergency

You are a captain of an engine company responding to the report of a car fire in a residential neighborhood. The homes in this area are old and often have a detached garage in the rear of the home, accessible via a shared driveway between two homes. As you approach the block, you see a column of black smoke rising from behind one of the homes. You pull up to the hydrant in front of the home. You have three firefighters with you. You run to the rear yard and see a fully involved car burning in a detached wood garage. There is a slight slope to the driveway, running down from the garage to the street. The homeowner is in the yard, ineffectually spraying water from his garden hose into the garage. As you call for a line, your truck company pulls up. It is staffed with a lieutenant and four firefighters.

Scenario Questions

1. What size line do you call for? Why?

2. What are your possible exposures?

3. Your firefighters are aggressively attacking this fire and are moving into the garage. Do you allow them to enter the garage? Do you insist on any precautions before they enter the garage? Why/why not?

4. The truck lieutenant is sending one of his firefighters up to the peaked garage roof to vent it. Do you agree with this tactic?

5. What if the garage is attached to a home? How would your tactics differ?

Answers

1. A, B, C, and D are all true.
   E. Approach the vehicle from the sides.
   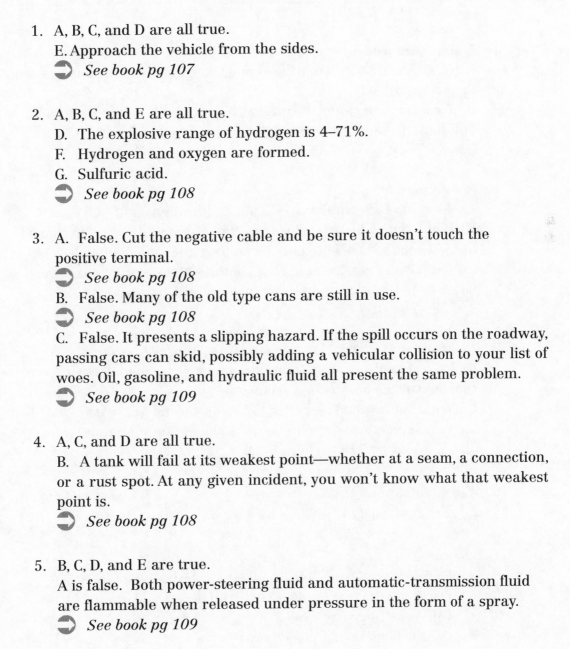 *See book pg 107*

2. A, B, C, and E are all true.
   D. The explosive range of hydrogen is 4–71%.
   F. Hydrogen and oxygen are formed.
   G. Sulfuric acid.
   See book pg 108

3. A. False. Cut the negative cable and be sure it doesn't touch the positive terminal.
   See book pg 108
   B. False. Many of the old type cans are still in use.
   See book pg 108
   C. False. It presents a slipping hazard. If the spill occurs on the roadway, passing cars can skid, possibly adding a vehicular collision to your list of woes. Oil, gasoline, and hydraulic fluid all present the same problem.
   See book pg 109

4. A, C, and D are all true.
   B. A tank will fail at its weakest point—whether at a seam, a connection, or a rust spot. At any given incident, you won't know what that weakest point is.
   See book pg 108

5. B, C, D, and E are true.
   A is false. Both power-steering fluid and automatic-transmission fluid are flammable when released under pressure in the form of a spray.
   See book pg 109

6. False. The vehicle may have obvious markings, or it may not. Even if it does, the markings may be obscured by smoke, flames, or darkness.
➲ *See book pg 110*

7. A and B are true.
C. False. The safest tactic may be to pull firefighters and civilians back out of the danger zone.
D. False. Arriving after the fire has begun, you'll have no way of knowing how long the tank has been exposed to heat or when it may fail.
➲ *See book pgs 110 and 111*

8. A and B are true.
C. Tests have shown that it will not ignite when shot.
D. While D may be true, more experience with this type of vehicle is needed before such a statement can be made authoritatively.
E. A CNG tank cannot BLEVE. It contains a gas, not a liquid. It can however fail, releasing the contained natural gas.
➲ *See book pg 111*

9. A. True
B. False. Use the reach of your stream at car fires to darken down and cool the vehicle before approaching it.
C. True, but it would be better to have a second apparatus block traffic for you. Remember, a car fire is not a single-unit operation.
➲ *See book pg 115*

10. A, B, and C are all true.
➲ *See book pg 115*

11. They are:
• Pierce the grill with a halligan tool or break a headlight.
• Pry up one side of the hood with a halligan tool.
• Penetrate the wheel well.
• Cut the hood with a power saw.
• Bounce the stream off of the ground below the engine so that it hits the fire in the engine compartment.
➲ *See book pg 116*

12. A, B, and D are true.
    C. Often the mechanism will have been damaged by the fire and won't be functional.
    ➥ *See book pg 116*

13. A. True.
    B. False. Insert the forked end over the hinge and bend it about 90°.
    ➥ *See book pg 116*

14. False. If the garage is attached to a dwelling, it may be necessary to stretch the line into the exposed dwelling to protect occupants and allow searches.
    ➥ *See book pg 118*

15. Using an exterior frontal attack from outside of the exterior door is generally safer for firefighters.
    ➥ *See book pg 119*

16. It should have enough slack to cover all parts of the building in order to be able to move quickly to cover any point of extension.
    ➥ *See book pg 119*

17. A, B, C, D, and E are all true.
    ➥ *See book pg 120*

18. A, B, and D are true.
    C is false. It will prevent the door from closing if the cables or springs fail. If the tracks fail, the whole door will come crashing down.
    ➥ *See book pg 120*

19. A, B, and C are true.

   D is False.

   • If the garage is attached with a bedroom over it, venting the roof won't vent the garage.

   • If the garage is attached to a dwelling, time may be better put to use stretching the attack line and searching for victims.

   • If the garage is detached, its roof rafters and deck may not be substantial or may be in disrepair and may fail early endangering the firefighters assigned to ventilate the roof.

   • Opening the garage door should provide ample ventilation.

   ⮕ *See book pg 121*

20. A. False. This should not be a single-unit response. Safety dictates the response of at least two units.

   ⮕ *See book pg 122*

   B. False. In some localities, the police may have that authority. In one instance, a fire chief was arrested for closing down the highway at the scene of an accident.

   ⮕ *See book pg 123*

   C. False. As soon as practical, chock the wheels whether the ground is level or sloped. If the car has a standard transmission and is in gear, the engine could start, causing the car to lurch forward and injuring firefighters in its path.

   ⮕ *See book pg 123*

21. A small spill can be covered by a commercial absorbent or a few shovels full of sand or dirt. For a larger spill, you may need to call a sanitation sander.

   ⮕ *See book pg 124*

Scenario Answers

1. A 1 ¾-in. line supplied with adequate water should be sufficient to extinguish the car fire and any extension to the structure. A larger line would be OK and would let you attack the fire from a greater distance.

2. The garage itself is likely to catch fire. You should be able to extinguish most of the interior fire from the outside of the structure if there is no ceiling below the roof rafters. If there is a ceiling over the car, it will have to be opened to expose any hidden fire above. Is there another garage next to or behind the involved one? How close to the house is the garage? If fire threatens to extend, you may need a second precautionary line. If there is an ember problem, you will need to check the surrounding area and structures for extension and, if found, have the resources to extinguish that fire. Brush and trees can ignite and then extend fire to surrounding structures. If the fuel line or fuel tank of the car fails, you might be faced with a flowing fuel fire. The slope of the driveway will allow the fuel to run toward the houses and the street. You will have to extinguish/contain the fuel spill.

3. Do not let your firefighters enter the garage right away. They should be able to knock the fire down from outside the garage. Once most of the fire is knocked down from the exterior, prop the garage door up with a pike pole or vise grip before letting anyone into the garage to complete extinguishment. Failure to prop up the roll down garage door could result in the springs or tracks failing as a result of the fire and the door crashing down on the firefighters or on the hoseline, cutting off their water supply and trapping them in the garage. Putting a pike pole in the track will prevent the door from coming down if the springs and/or cable fail. If the tracks fail, the entire door will still fall.

4. There is already plenty of ventilation available from the opened garage door. Garage roofs may not be as sturdy as a residence roof and may not be well maintained. Venting the peaked roof will be time-consuming, and the additional ventilation will not be worth the risk involved. Countermand the lieutenant's order.

5. If it is an attached garage, the first line might have to be stretched into the dwelling to prevent extension into the living space and to protect firefighters as they search for victims and extension. This first line can attack the fire from an interior entrance to the garage door or merely remain in position as a precaution against extension while a second line attacks the fire from the exterior garage door.

6

Kitchen Fires

Questions

1. Studies were done showing that different food products react differently when overheated in a pot or pan on a stovetop burner. Which of the following statements regarding these studies is true?

   A. Corn oil started to smoke in 10 min and was ignited in 22 min.

   B. When the corn oil ignited, the flames reached a height of 8 ft in 2 min.

   C. Beans, despite being heated to 303°F, did not ignite.

   D. The beans gave off more smoke than the corn oil.

2. You are called to a building for a smoke condition in the hall. It smells like burnt food. You can expect the occupant of the apartment who has burned his dinner to readily admit he is responsible for the smoke condition. True or false?

3. You are called to a food-on-the-stove incident. If no one answers the door, it will be safer and easier to enter through the window rather than forcing the door. True or false?

4. **We are sometimes called to a home because the occupant says that he is locked out and that he has left the stove on. We are expected to force entry and let the occupant into the residence. Which of the following statements about these types of calls are stated *incorrectly*?**

   A.   Sometimes the caller reports that a small child is locked inside the apartment to ensure our prompt response.

   B.   On gaining entry, you may find no pot, no burning food, and no small child.

   C.   The caller may have stated that the stove was on to make the call seem more serious and to ensure our response and assistance.

   D.   These are routine runs, and we simply let the occupant back into the residence.

5. **Which of the following statements about food-on-the-stove incidents is stated *incorrectly*?**

   A.   A light haze might indicate a less serious incident.

   B.   Heat from a window would indicate fire extending to combustible furnishings or structural elements.

   C.   If no one responds to your banging on the door, assume no one is at home.

   D.   A car in the garage or in the driveway or toys on the porch might indicate someone is home.

6. **You respond to a multifamily building for a food-on-the-stove incident, and there is heavy smoke coming from the apartment window. You should remove occupants from the stairs and public hall before you open the door. True or false?**

7. **You can usually vent the hallway and stairs of a multiple dwelling by opening the bulkhead, scuttle, and/or stairway windows. This can be done for high-rise buildings also. True or false?**

8. **What harmful and odorous substances are typically produced by food-on-the-stove incidents?**

9. **A pan on the stove can reach ___?___ °F in less than 10 min and can cause serious injury if grabbed with an ungloved hand. Fill in the blank.**

10. **What are some hazardous substances commonly stored in the vicinity of the stove?**

11. **Which of the following statements are true?**

    A. At a food-on-the-stove incident with no fire extension and a medium smoke condition, you should order your firefighters to break the windows to vent the apartment.

    B. Opening double sash windows two-thirds at the top and one-third at the bottom will amply serve to clear the smoke from the apartment.

12. **The new energy-efficient windows can not be easily vented and often must be broken to ventilate them. True or false?**

13. **Smoke generated at a food-on-the-stove incident is hot and will rise and vent quickly once a window is opened. True or false?**

14. **At a food-on-the-stove incident, you will have to find the kitchen and locate the stove. Which statement about its location is *incorrect*?**

    A.   In private homes, the kitchen is often located in the rear of the house on the first floor.

    B.   In New York's so-called railroad flats, the kitchen is found in the rear.

    C.   It is rare to find more than one stove in an occupancy.

    D.   In newer apartment buildings, the location of the kitchen is usually the same from floor to floor.

15. **You are called to a food-on-the-stove incident. It is the result of oil burning in a pan on the stove. Which of the following tactics do you approve of?**

    A.   First the firefighter shuts off the stovetop burner.

    B.   Your firefighter takes the burning oil and places it in the sink. He then turns on the faucet to extinguish it.

    C.   A firefighter covers the pot to extinguish the fire and then places it in the sink and allows it to cool naturally.

    D.   The pot continues to generate smoke after the flame is extinguished by placing the cover on it, so a firefighter removes it from the building.

16. **If you arrive at a food-on-the-stove incident and find a smoking pot, you need not search for fire extension since there are no flames. True or false?**

17. **You are dispatched to a call for smoke in the hallway from apartment 6C on the 6th floor and another call for food-on-the-stove from apartment 4A in the same building. When you arrive, you find food burning on the stove in apartment 4A. Why must you investigate the call from apartment 6C?**

18. **The following are statements about tactics for a stove fire that has extended beyond the stove. Which of the following are stated *correctly*?**

   A. Any damage to the entrance door is inconsequential since delay could result in a loss of life or even more structural damage.

   B. Check for fire extension on the floor above and in the adjoining apartment at the kitchen pipe chase.

   C. If it is a top floor fire, check the cockloft.

   D. This is a structural fire, and you will have to break windows to ventilate it.

19. **You may have to do substantial damage to the structure if you think fire has extended to it. The following are statements about overhauling a kitchen fire. Which are stated *incorrectly*?**

   A. For a fire in a built-in wall oven, you may have to check around all sides of the oven for extension.

   B. Scrape the char off of the wall and cabinets with the adze end of your halligan or use your ax.

   C. If you have to open the ceiling to check for fire, check first where the ceiling is already pierced, since this is where it is easiest for fire to enter.

   D. If there is a range hood over the stove, there will be ductwork that must be checked.

   E. You may have to move the stove to check behind it for extension.

20. **Microwave ovens are sometimes the cause of a kitchen fire. Which of the following statements about microwave ovens are stated *incorrectly*?**

   A. Although microwave ovens heat without flames, they have been known to cause fires and to generate smoke.

   B. A paper bag with a staple embedded in it can ignite when heated in a microwave.

   C. Aluminum foil can cause arcing and fire in a microwave.

   D. There have been reports of microwaves spontaneously turning on as a result of electrical storms.

Topics for Drill

1. Identify the typical location of the kitchen in the private dwellings and multiple dwellings in your district. Is there an exterior clue that you can use to locate the kitchen? (Window size, vent fan, door?) Take photos of these buildings and use them at drill to familiarize your firefighters with the buildings. If you are computer literate, these photos can be put into Microsoft PowerPoint® and annotated with text to facilitate the learning process.

2. Review your venting policy with your firefighters. Explain how to vent without causing damage and when you expect them to vent without regard to damage. Having everyone on the same page here will go a long way to making your fireground operation professional and promoting good customer relations with the people you serve.

3. Discuss your policy regarding forcing entry at food-on-the-stove incidents. Include the necessity to verify the occupants' identity before letting them into the dwelling.

    A. What alternatives do you have to forcing the entry door when you respond to a food-on-the-stove incident?

    B. What hazards do the firefighters face when utilizing an alternative entry method?

Scenario: Kitchen Fire Emergency

You are the captain of a ladder company, riding with four firefighters in addition to yourself. You are called to an apartment house for an odor of smoke. The report states that there is smoke on the top floor of what turns out to be a three-story woodframe apartment building with four apartments per floor. You arrive first, with the chief and an engine company right behind you. The engine rides with an officer and three firefighters.

There is no smoke showing and you notice the building has been retrofitted with replacement windows. When you enter the building, you notice the familiar smell of burnt food and announce to the chief, "It smells like food." You have three of your firefighters start to knock on doors looking for the source of the odor.

People on the floors above are looking down the open stairway and asking what is going on and where is the fire.

Scenario Questions

1. What action will you take if you check all of the doors on the 4th floor but no one answers the door?

2. Should the engine company stretch a line?

3. When should you notify the chief that the situation is under control?

4. If you find the wall and cabinets scorched and a smoldering pot on the stove, what actions would you take?

5. Once you gain entry to the apartment, firefighters find a pot of burning oil on the stove. They turn the stovetop burner off and extinguish the burning oil with the water extinguisher. Do you agree with this tactic?

Answers

1. A, B, and C are true.

   D. Corn oil gave off more smoke.

   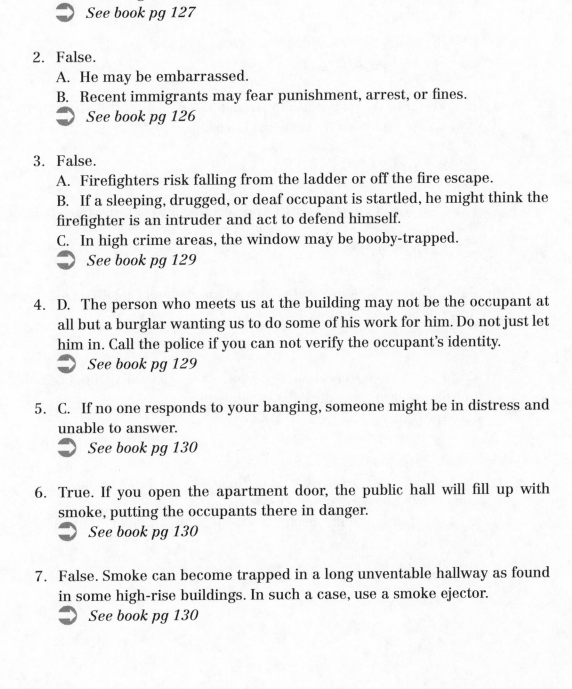 *See book pg 127*

2. False.

   A. He may be embarrassed.

   B. Recent immigrants may fear punishment, arrest, or fines.

   See book pg 126

3. False.

   A. Firefighters risk falling from the ladder or off the fire escape.

   B. If a sleeping, drugged, or deaf occupant is startled, he might think the firefighter is an intruder and act to defend himself.

   C. In high crime areas, the window may be booby-trapped.

   See book pg 129

4. D. The person who meets us at the building may not be the occupant at all but a burglar wanting us to do some of his work for him. Do not just let him in. Call the police if you can not verify the occupant's identity.

   See book pg 129

5. C. If no one responds to your banging, someone might be in distress and unable to answer.

   See book pg 130

6. True. If you open the apartment door, the public hall will fill up with smoke, putting the occupants there in danger.

   See book pg 130

7. False. Smoke can become trapped in a long unventable hallway as found in some high-rise buildings. In such a case, use a smoke ejector.

   See book pg 130

8. Sulfur, nitrogen dioxide, and carbon monoxide.
   ➲ *See book pg 131*

9. 300°F.
   ➲ *See book pg 131*

10. Bleach, ammonia, pesticides, oven cleaner, and drain cleaner.
    ➲ *See book pg 131*

11. A. False. Breaking glass isn't warranted for food-on-the–stove incidents that do not escalate into structural fires.
    B. True.
    ➲ *See book pg 131*

12. False. They can often be partially or completely removed by detaching them from their tracks.
    ➲ *See book pg 131*

13. False. While this smoke will be initially hot, it will cool down quickly and may not be easily vented.
    ➲ *See book pg 132*

14. C. In some areas, it is common to have a second stove in the basement to help prepare large meals at holiday time.
    ➲ *See book pg 132*

15. A and C are good tactics.
    B. Adding water to burning oil could cause the flame to flare up, endangering the firefighter. This is not a good tactic.
    D. A good idea, but be careful not to spill the hot oil on yourself or anyone else.
    ➲ *See book pg 132*

16. False. There may have been active flame before you arrived. You must check for extension.
    ⮕ *See book pg 133*

17. Apartment 6C may be reporting smoke from a different fire.
    ⮕ *See book pg 134*

18. A, B, and C are stated correctly.
    D is incorrect. If the extension is truly minor, you may not need to break any windows. If there is substantial smoke and heat, vent as you would at any structural fire.
    ⮕ *See book pg 135*

19. D is incorrect. Many range hoods aren't connected to ductwork.
    ⮕ *See book pg 137*

20. A, B, C, and D are correct.
    ⮕ *See book pg 138*

Scenario Answers

1. A. You must check all of the doors in the building. The smell might be coming from any apartment in the building. The stairs are open, and there is nothing to restrict the odor from permeating the entire building.

    B. If no one answers at an apartment, push in on the door to open it as much as it will give and smell by the crack of the door. The one with the greatest odor is probably the one with the burning food.

    C. Take a close look at the windows from the outside of the building. Look for signs of smoke venting from an open window or a haze inside the closed windows. Remember, these replacement windows seal the building up pretty tightly and you might easily miss smoke behind the window.

    D. If you find an apartment with smoke venting from a window, you can consider forcing a window and entering via a ladder. Make plenty of noise before entering and remember that you may startle the occupant who may have been asleep or drugged and think that you are an intruder.

    E. If you do not get into the apartment, a simple food-on-the-stove incident might escalate into a structural fire. Also, the occupant might be unconscious and unable to respond to your knocking. Once you determine that you have a smoke condition and not just an odor, you must find and enter the apartment.

2. Stretching a line for an odor is not usually necessary. Stretching a line for a smoke condition is. For a light smoke condition, the engine can stretch a precautionary line to the front door. It can be charged if you determine that it is necessary.

3. No matter how minor the incident seems, do not notify the chief that you have the incident under control until you have found the source of the odor or smoke, entered the apartment, made a search and determined that all is well. Keep him informed of your progress. Let him know when you locate the apartment. Let him know when you gain entry and let him know the final outcome.

4. A. Open the cabinets and see if anything inside is smoldering or burning.
    B. Scrape the char off of the cabinets to see if it is still smoldering.
    C. You may have to check behind the cabinet for fire extension. This will require prying the cabinet away from the wall.
    D. Open the ceiling and soffit if it is possible that fire has extended to it.
    E. Depending on the amount of extension, check the floor above and adjoining apartment for fire extension. If the building is balloon construction and fire has entered the exterior wall, check the floor below.
    F. Wet the charred cabinets and wall with your water extinguisher or hose line. Don't use more water than is needed. Remember, people still have to live here.
    G. If there was more than minor extension, charge the line as a precaution as you open up and search for extension.

5. It is not a good idea. Putting water onto the burning oil can result in the oil fire flaring up. It would be better to simply cover the pot. This would extinguish the oil by excluding oxygen.

7

Mattress Fires

Questions

1. Which of the following statements are *incorrect*?

    A.   After a mattress fire, you must check the adjoining apartments and the apartment above.

    B.   Someone with a heart condition, asthma, or other pulmonary condition might be adversely affected by just a little smoke.

    C.   For a smoldering mattress that did not reach active flaming stage, only minimal overhaul of the bedroom is needed.

    D.   The ceiling will have to be pulled even if the mattress fire did not reach the active flaming stage.

2. Which of the following statements are accurate?

    A.   If flame and heated gases were produced by a mattress fire, it is important to check to see whether the fire penetrated the ceiling or walls.

    B.   Answer A, plus check ceiling light fixtures and pipes that penetrate the ceiling, electrical outlets, switches, and other openings that might allow fire to extend behind the plaster.

    C.   Intact plaster or plasterboard ceiling or wall rarely halts the spread of fire.

    D.   If the ceiling is damaged from the heat, you must check it for fire.

    E.   If you find charred wood as you open the ceiling, open as far as the next bay.

    F.   Pulling the ceiling may cover a victim with debris.

3. **If a wall is warm, open it up and check for fire behind it. True or false?**

4. **Which of the following statements are accurate?**

    A. If you find fire extension the wall, close to the floor, quickly wet it down and be thankful that it did not extend any farther.

    B. You find fire extending up the wall near the ceiling when you open up the wall. Next, you should open up the ceiling and check the floor above for extension.

5. **Which of the following statements about overhaul are true?**

    A. Check around wall switches and outlets and open them up if you suspect fire has extended into them.

    B. If the building is a platform-frame building, consider the possibility that if fire penetrated the wall, it might have dropped to the floors below or extended to the floor above, even into the attic or cockloft.

    C. Open closets and drawers to make a thorough examination.

    D. Remove burned or smoldering items from closets or drawers and thoroughly soak them with water.

    E. Clothing, money, jewelry, and sentimental items can and should be saved.

6. **Thoroughly soaking the surface of the mattress with water from a hand line will extinguish mattress fires. True or false?**

7. **Which of the following statements concerning fires in foam mattresses are true?**

    A. Since foam mattresses do not support burrowing fires, they are quickly extinguished with a small amount of water.

    B. If a foam mattress is not sufficiently cooled by water, it will continue to generate hot, combustible gases that can collect around the mattress and even fill the room.

    C. A dry chemical extinguisher should be used to extinguish these fires.

    D. Thoroughly soaking the mattress will sufficiently cool the mattress and stop the generation of flammable gases.

8. **Once you have cut open a mattress and wet it down, there is no need to remove it from the building. True or false?**

9. **Which of the following statements are accurate?**

    A. Firefighters removing a burnt mattress from the building might be injured if the mattress flares up as they are transporting it.

    B. As firefighters move the mattress, they can create a bellows effect causing air to be sucked into the mattress. This can cause any smoldering fire within the mattress to flare up.

    C. Don't bother trying to fold the mattress before moving it. This might actually cause more air movement and result in the mattress flaring up.

    D. Fold the mattress so that the burned side is on the outside and tie the mattress in place with a short length of rope before moving it.

10. **List three precautions that you should take before throwing a burnt mattress out of a window.**

11. **Which of the following statements about moving a burnt mattress out of a building are true?**

    A. If you are moving a burnt mattress out of the building, via the stairs, you should have a charged line in the street ready to thoroughly douse it with water once it is out of the building.

    B. If possible, avoid using the elevator to remove the burnt mattress.

    C. If it is absolutely essential to use the elevator, take one or more extinguishers along with you.

    D. At the first sign of flame or smoldering fire, immediately douse it with water from the extinguisher.

    E. Do not stop the elevator until you are on the ground floor.

12. **Which of the following precautions for fires in stuffed furniture are *incorrect*?**

   A.   The same precautions that you would take for a mattress fire also generally apply to fires in stuffed furniture.

   B.   It is unlikely that smoldering embers will go unnoticed in an overstuffed chair.

   C.   If fire has involved an overstuffed chair or couch, remove it from the building even if the burned area is superficial.

   D.   If only the cushion was involved, you can soak the cushion in a tub or the sink.

Topics for Drill

1. Ask how your firefighters would remove a burnt mattress form a building. How would this task vary for the different building types in your district?

2. What hazards or physical obstructions would make the task of removing the mattress difficult or dangerous?

3. What precautions should be taken to prevent injury to firefighters when removing a burnt mattress from the building?

4. Ask your firefighters to explain why a water extinguisher may not contain enough water to safely extinguish a mattress fire.

Scenario: Mattress Fire Emergency

You receive a call at 0400 hours on Saturday morning. It is winter and all of the buildings are closed up tightly against the bone-chilling cold. The call is for smoke coming out of a bedroom window in a two-story private dwelling. Your response is one engine staffed with three firefighters and an officer and one ladder company staffed the same and a chief. You are the truck officer and you arrive first. Upon arrival, you see smoke coming out of the front bedroom window on the 2nd floor. You have to force entry into the front door, and as you go up the interior stairs, you see smoke puffing from underneath a bedroom door. Your engine company is on the scene and you call for a line. You enter the bedroom, make a search, and find no one in the room. Flames can be seen through the smoke. The mattress is burning.

Scenario Questions

1. One of your firefighters has the 2 ½-gal water extinguisher and you direct him to play the stream on the mattress. He knocks the fire down, and you announce to the chief that the fire is extinguished and that hoseline will not be needed.

    A. What hazards could be present if the mattress is a foam mattress?

    B. What hazards could be present if the mattress is a traditional spring and stuffing mattress?

    C. Should a line usually be stretched at mattress fires?

2. Ask your firefighters:

    A.   When would they leave the burnt mattress in the room after extinguishing all visible fire?

    B.   How should you remove a mattress to the street and what precautions should you take for each method?

Answers

1. B and C are correct.

   A. You must decide at each incident how far you should extend your search.

   D. If there was no active flaming, the ceiling will not have to be pulled.

   ⮕ *See book pg 142*

2. A, B, D, and F are true.

   C. Intact plaster can halt the spread of fire.

   E. Open the ceiling until clean wood is revealed.

   ⮕ *See book pgs 142 and 143*

3. False. If it is too hot to hold your hand on, you must open it up to check for fire.

   ⮕ *See book pg 143*

4. B is true.

   A. You should next open the bay near the ceiling to check for extending fire.

   ⮕ *See book pg 143*

5. A, C, D, and E are true.

   B. This is true of balloon frame, not platform frame. Platform frame may limit vertical extension.

   ⮕ *See book pg 143*

6. False. A stuffed mattress must be cut open and water must be applied to the interior. Fire can burrow into a mattress and smolder there, protected from any water that you apply to the surface. (This is not true of foam mattresses.)

   ⮕ *See book pg 143*

7. B and D are true.

   A. False. You must use sufficient water to thoroughly cool the foam mattress.

   C. False. Dry chemical will not cool the mattress nor suppress the production of flammable gases.

   ⮕ *See book pg 144*

8. False. You might still have missed smoldering embers in the stuffing. Remove the burned mattress from the building as a precaution to ensure that no one is harmed should it flare up later on.

   ⮕ *See book pg 144*

9. A and B are true.

   C. False. Folding the mattress will prevent air from being pumped inside the mattress by the bellows effect.

   D. False. Fold the mattress in half so that the burned side is within the fold and tie in place with a short length of rope before moving it.

   ⮕ *See book pgs 144 and 145*

10. 1) Make sure that no one is standing below the window.
    2) Check to see that no victim has jumped and is lying below the window.
    3) Assign a firefighter to inspect the area and keep everyone clear.

    ⮕ *See book pg 145*

11. B, C, and E are true.

    A. False. Have a line standing by in case the mattress flares up in the building as it is being moved.

    D. False. If the mattress ignites, stop the elevator at the nearest floor to provide a means of egress should the extinguishers not be able to control the fire.

    ⮕ *See book pg 145*

12. A and D are correct.

    B. Incorrect. The danger of smoldering embers going unnoticed in an overstuffed chair is a very real threat.

    C. Incorrect. If the burned area is small, cut it out and soak it in a bucket.

    ⮕ *See book pg 146*

Scenario Answers

1. Keep the following in mind:

    A. A foam mattress will continue to generate flammable gases unless it is thoroughly cooled. These gases can ignite under the right conditions, endangering firefighters working in the fire room.

    B. Fire burrows into traditional spring mattresses where it can smolder unseen. It is necessary to cut open and thoroughly soak these mattresses to prevent reigniting. When moving these mattresses, air is pumped into the interior of the mattress. Firefighters have been burned when the mattress they were moving suddenly flared up.

    C. The water extinguisher may not fully extinguish the fire and it may flare up as it is being moved. In the case of a foam mattress, flammable vapors could ignite and cause a flare-up. Stretch a hoseline and thoroughly soak the mattress as needed.

2. Some of their answers can include the following:

    A. If there is only a truly minor burn in a mattress and it has been cut open and checked for interior burning and the burned area has been cut out and thoroughly soaked, it would be OK to leave the mattress in the building. When in doubt, take it out.

    B. Before moving a mattress, fold it in half if possible and tie it so that it does not unfold. Then you can toss it out the window if it is safe and practical to do so. Make sure to check the area below before you toss the mattress out and post a safety man to keep people out from under the window. If you can't toss it out the window, then take it down the stairs. Be sure to have a charged line standing by in case it lights up as it is being carried down the stairs. If you must put it in an elevator to remove it from a building, bring a few water extinguishers with you in case it ignites. It would be safer not to put it in an elevator if at all possible.

8

Trash Fires

Questions

1. A standard 2½-gal water extinguisher can project a stream of water how many feet?

2. Which of the following statements about rubbish fires are stated *incorrectly*?

   A. If rubber, carpeting, or upholstery is burning, hydrogen cyanide could be present in the smoke.

   B. Acrolein is produced by burning carpeting.

   C. Burning plastic may result in hydrogen chloride gas being present in the smoke. Phosgene could be present if PVC is burning.

   D. Burning wood generates relatively safe smoke.

3. How can you protect yourself from the dangers of smoke generated by a trash fire?

4. Our prime directive is to save life. What should be the primary consideration at trash fires on highways?

5. **We must protect firefighters when they are operating at highway trash fires. Which of the following statements regarding this is *not* true?**

   A. Position your apparatus between firefighters and oncoming flow of traffic.

   B. Only bring one vehicle onto the highway to reduce your exposure.

   C. Place flares.

   D. Use the police to control the flow of traffic.

   E. Get the job done quickly and get the firefighters out of harm's way.

6. **There is no life hazard at a rubbish fire. True or false?**

7. **When trying to stay out of the smoke generated by a burning rubbish fire (by keeping the wind at our backs) what can unexpectedly place us back in the smoke?**

8. **Which of the following statements about hose selection for rubbish fires is stated *incorrectly*?**

   A. Whichever line you choose must have enough reach to keep you away from rubbish while delivering enough water to quickly extinguish the flames.

   B. Always stretch a hoseline—even for a small rubbish fire.

   C. A booster or trash line may be adequate for some fires, but for larger fires or more hazardous ones, a 1 ¾-in. or 2 ½-in. hoseline may be required.

   D. At a large, deep-seated fire in a garbage dump, using a tower ladder stream will keep your firefighters out of harm's way.

9. **Which of the following are hazards you are exposed to when manually overhauling a large area rubbish fire?**

   A.   You risk a puncture injury to your feet and legs.

   B.   A sprain or fall.

   C.   The chance of being injured by an exploding container.

   D.   Exposure to hazardous material.

10. **How can you avoid the hazards listed in question 9?**

11. **Which of the following statements about dumpster fires is stated *incorrectly*?**

   A.   Dumpster fires present certain hazards to firefighters. If the dumpster is located on hospital property, it might contain illegally dumped medical waste. Such waste will be easily recognized at a dumpster fire, because it will be in red plastic bio-waste garbage bags.

   B.   There is no telling what a metal dumpster might contain.

   C.   The type of occupancy served by a given dumpster will tell you what is in the dumpster.

   D.   Large trash dumpsters are found at construction sites, commercial occupancies, buildings under renovation, and anywhere that large amounts of disposable rubbish are generated.

12. **Indicate which of the following statements are true.**

   A.   Enter a dumpster cautiously when overhauling.

   B.   When possible, overhaul the contents hydraulically.

   C.   Since there is no life hazard at a dumpster fire, search is never needed.

13. **Which of the following exposure considerations at dumpster fires are correct?**

   A. Dumpsters are often positioned near a building.

   B. Position your hand line so that it can apply water to the fire.

   C. You may have to conduct an examination of the building's interior if fire extension is a possibility.

   D. One way to prevent extension of fire from a large dumpster fire is to roll it away from the building. This can be done by one or two firefighters.

14. **We can expect to be called quickly to a junkyard fire caused by a careless torch operator cutting apart a junk car. True or false?**

15. **You respond to a junkyard at night, and in the yard, you find a boat fully involved in fire. It is isolated and there are no exposure problems. You must make an aggressive attempt to search the boat for any possible victims. True or false?**

16. **Which of the following statements regarding overhauling at junkyards are true?**

   A. Use a minimum of staffing but don't allow firefighters to move heavy objects alone.

   B. A burrowing fire in a large pile of granular material may continue to burn even after repeated attempts at extinguishment. Such a pile will require firefighters with shovels to dig into the pile to locate the burning material.

   C. If heavy digging equipment is needed but not available in the middle of the night, hydraulically overhaul the material and then cover it with protein foam to prevent flare up until the next day when the required equipment becomes available.

   D. A burning pile of autos may have to be systematically taken apart, extinguished, and then relocated into a new pile.

17. **Vacant buildings, unless properly sealed, can become large garbage pails. Which of the following statements about rubbish fires in such buildings are *incorrect*?**

    A. There is no life hazard other than the firefighter in such buildings.

    B. These buildings are usually structurally sound and will only collapse after serious fire damage.

    C. Serious potential for firefighter injury exists in such buildings.

    D. The presence of clothes, bedding, or food or an accessible opening in an otherwise sealed building may indicate that it is inhabited.

    E. These buildings can be used as drug dens.

18. **Which of the following statements about rubbish fires in abandoned buildings are stated *incorrectly*?**

    A. When possible, extinguish the fire from outside. Use the reach of your hose stream to keep you out of the building.

    B. Use portable lights to illuminate the area.

    C. Shine lights down through holes in the floor and stairways to prevent firefighters from walking into them.

    D. Holes in the floor can be made safe by covering them with doors taken from adjoining rooms.

    E. Enter the room with the burning rubbish and get close to the fire before opening your hoseline.

    F. Often you can easily spot existing holes in the floor from the floor below the fire so assign someone to check out the floor below before you move in on the fire.

    G. Holes in the floor, ceiling, and walls, as well as missing doors and strewn rubbish can allow fire to spread with alarming speed.

19. **Why is it important to search the floors above and below firefighters working in an abandoned building?**

20. Which of the following statements about operating in an abandoned building are stated *incorrectly*?

A. If firefighters are operating above grade, raise ladders to provide them with an alternate means of egress.

B. Do not ventilate sealed windows. Opening them will provide air to the fire and cause it to accelerate.

C. When deemed safe, enter the building with masks and full firefighting gear. Maximum supervision and a large commitment of personnel can reduce the possibility of firefighter injury.

D. After the bulk of the fire has been extinguished by an exterior stream and before committing firefighters to interior operations, a senior chief or officer should reevaluate the building to determine whether it's safe to enter.

E. Firefighting shouldn't be conducted in any derelict building without adequate resources standing by to assist in an emergency.

F. Elevated streams can deliver as much as 1,000 gpm. Water pouring out of an abandoned building after elevated stream use is a danger sign of collapse.

21. The author explains a marking system for vacant buildings. Which of the following is stated *incorrectly* regarding that marking system?

A. Paint an 18×18-in. square over the entrance door to indicate that no unusual hazards are present in the vacant building.

B. If structural damage or interior hazards are present, paint an additional line within the square, going on a diagonal from the upper right-hand corner to the lower left-hand corner.

C. The mark described in B means that an examination must be made before interior operations can be conducted.

D. A square with a full X painted inside means that severe interior hazards or structural damage exists and that all operations should be conducted from the exterior.

E. If interior operations are considered in a building marked with an X, the officer in command must approve it, an interior survey must be conducted, and the operations must be carried out by a minimum number of personnel.

22. **If an outside rubbish fire exposes a structure, extinguish the fire as quickly as possible to reduce the likelihood of the fire extending to the building. Which statements about these fires are *incorrect*?**

A.   If you suspect extension, stretch a second, precautionary hoseline to the interior as soon as possible, even before you've extinguished the rubbish fire.

B.   The examination for extension can start on the outside. Look for cracked windows and look through windows for signs of fire or smoke inside.

C.   If no occupant is on hand and if possible, wait until someone with a key arrives before forcing entry to minimize damage to the structure.

D.   If smoke is pushing out of the siding, check to see if the underlying framing or insulation ignited.

Topics for Drill

1. Identify vacant buildings in your area and survey them to see if they are accumulating flammable rubbish. Create a preplan for how are you going to fight fires in these buildings and the precautions you will take to ensure no one is injured.

2. Do you have a junkyard in your area? Visit it and identify the hazards that you might face if called to fight a fire in the junkyard.

3. Take your firefighters out and identify the location of dumpsters in your area and try and determine what hazards are typically found in them. If you are called to a fire in the dumpster, is extension to a structure likely? If so, how will you prevent extension? Discuss with your firefighters what wastes you are likely to find in the dumpsters in your area and the danger they pose to firefighters. Have the firefighters explain how they will avoid these hazards at a dumpster fire.

4. Do you have illegal dumpsites in your area? Visit them and determine if hazardous waste is present. What dangers would your firefighters face if called to fight a fire in the trash? Try and get the appropriate agency or individual to clean up the trash before you are called to the site for a fire.

Scenario: Trash Fire Emergency

You are dispatched as the only engine responding to the report of an outside rubbish fire in the rear of an auto dealership in your area. The dealer both sells and services cars. You ride with an officer and three firefighters, and you are the lieutenant.

When you arrive, you encounter a rubbish fire in a large dumpster located close to the roll-down door at the rear of the dealership. Flames are rising 10 ft into the air, and the wind is pushing them against the roll-down door. There is a hydrant 50 ft from the dumpster.

Scenario Questions

1. List the tasks that must be performed to successfully and safely fight this fire.

2. Do you have adequate resources to successfully and safely fight this fire?

3. Is it possible that fire has already extended into the dealership?

4. How would you determine if fire had extended into the dealership?

5. How would you apply water to the fire and why?

Answers

1. 20 ft or more.
   ⮕ *See book pg 149*

2. A, B, and C are correct.
   D. Burning wood can give off formaldehyde. Creosote will be in the smoke of burning wood if it has been treated with creosote.
   ⮕ *See book pg 150*

3. Stay out of the smoke or wear SCBA.
   ⮕ *See book pg 151*

4. Protect firefighters and fire apparatus from being struck by moving vehicles.
   ⮕ *See book pg 151*

5. A, C, D, and E are true.
   B. Position a second apparatus to shield firefighters from oncoming traffic. This should be done at all roadway incidents as a precaution.
   ⮕ *See book pg 151*

6. False. Firefighters are at risk.
   ⮕ *See book pg 151*

7. Variable wind.
   ⮕ *See book pg 152*

8. B. For a small fire, a 2 ½-gal water extinguisher may be enough.
   D. This is true if you remember to keep the firefighters in the basket out of the smoke or suited up in SCBA.
   ⮕ *See book pg 152*

9. All answers are correct.
   ➲ *See book pg 152*

10. Consider using your deck gun or an aerial stream to hydraulically overhaul a large area or a high pile of rubbish.
    ➲ *See book pg 152*

11. A. The color of the bag may be obscured by the smoke and not recognized until the fire has been extinguished.
    C. Although the type of occupancy gives you a clue as to what to expect, any dumpster is susceptible to illegal dumping of hazardous materials.
    ➲ *See book pg 152.*

12. A. False. Do not enter the dumpster.
    B. True.
    C. False. It's unlikely—but not impossible—that there is a life hazard in the dumpster. Playing children or foraging vagrants might be trapped or injured.
    ➲ *See book pg 154*

13. A and C are correct as stated.
    B. Position your line so you can alternately wet down the exposure and apply water to the fire.
    D. This is only true for small dumpsters.
    ➲ *See book pg 154*

14. False. They typically call us only when they fail to extinguish it themselves and the fire spreads.
    ➲ *See book pg 154*

15. False. *Fully involved* are the key words. Anyone onboard is dead. Again, you are the life hazard, and a cautious approach is indicated.
    ➲ *See book pg 155*

16. A, C, and D are true.
    B. Don't place firefighters on top of the pile; get heavy digging equipment.
    ⟳ *See book pg 159*

17. C, D, and E are true.
    A. Children play in these buildings and vagrants live in them. There may be a life hazard other than firefighters.
    B. Such buildings, if left vacant for a prolonged period of time, can suffer structural damage from the elements. Beams, floors and roofs will rot. Repeated fires will further weaken the structure, and accumulated trash will crate a heavy fire load, absorb water, and add weight to the floors.
    ⟳ *See book pg 159*

18. A, B, D, F, and G are accurate.
    C. Shine the lights up through the holes in the floor.
    E. Extinguish as much fire as possible from the doorway before entering the room.
    ⟳ *See book pg 160*

19. To prevent a rekindle after firefighters leave the scene and to protect firefighters from becoming entrapped by extending fire. It is possible that more than one fire has been set. It is also possible that the fire has extended to the floor above or below.
    ⟳ *See book pg 160*

20. A, D, and E are correct.
    B. If windows are sealed, open them for ventilation and as a ready escape route.
    C. Maximum supervision and a minimum commitment of personnel can reduce the possibility of firefighter injury.
    F. Runoff is a good sign. If it does not run off, the water will add tremendous weight to the structure and could cause collapse.
    ⟳ *See book pg 160*

21. A. The 18×18-in. square indicates that no unusual hazards were present at the time of marking. Hazards may have developed since then.
    ⟹ *See book pg 161*

22. A, B, and D are correct.
    C. Whatever damage you do to the building in gaining entry will be small compared with what hidden fire could do if not quickly found and extinguished.
    ⟹ *See book pg 162*

Scenario Answers

1. Task List

   • Establish a water supply.
   • Apply water to the fire and the exposed building.
   • Check for extension into the building.
   • Extinguish any extension and search for any victims.
   • Overhaul.
   • Provide security for the structure if you have forced entry.

2. This fire has the potential to have already extended to the structure. You should be simultaneously applying water and checking the exposure. You do not have the manpower to do this. You will need at least a truck company to assist you. If the fire has extended to the building or if you suspect it has or will, you will need a full alarm for a structural fire. A line will have to be stretched to the interior, and the interior will have to be searched. It may be wise to get the full alarm right away, just in case fire has extended into the building. You can always send them home if you don't need them.

3. The flames impinge on the metal roll-down door. This alone could cause fire extension to any combustible in the building in contact with the door. It is also possible that the flame or heat has ignited the framing around the door.

4. A quick look in the windows may reveal that smoke has entered the building. If smoke is in the building, definitely transmit an alarm for a structural fire. If it is only smoke and no fire is found, you can always send the companies back. The lack of obvious smoke does not necessarily mean that fire has not extended.

   The existing windows may not give you a view of the area near the roll-down door or there may be a small spot of smoldering fire inside that is not creating much smoke. You will have to force entry into the dealership and physically check for fire extension. Your thermal imaging camera will be of assistance in looking for hidden hot spots in the wall and around the door. Another thing to consider is whether embers rose up in the heat plume from the burning dumpster falling onto the roofing and igniting it.

5. You could quickly douse the flames with your deck gun while you stretch a supply line to the hydrant. You must consider, however, the possibility of driving the flames into the building. In this case, it might be better to stretch a hand line and position it so that you can apply water to the door and frame as well as the dumpster. This follows the theory of putting the line between the fire and the exposure. You might be able to prevent extension with your tank water while your supply line is being stretched. Once you have a water supply, you can use your deck gun to complete extinguishment.

Part II
Carbon Monoxide

9

The New Response

Questions

1. In the book, the author mentions a number of organizations that are interested in the CO problem. How many of them can you name?

2. All of the groups mentioned in this chapter stand to gain or lose time, money, or prestige as a result of the rapidly changing CO landscape. True or false?

3. What groups or agencies are involved in setting standards related to CO and CO detectors?

4. There are two standards that pertain to home CO detectors. One is UL 2034 and the other is IAS U.S. Requirement no. 696. Which standard represents an attempt of the American Gas Association and Canadian Gas Association to reduce low-level activations of existing alarms and to improve the reliability of new home detectors?

5. **CO has been around as long as fire has, but only recently has it been given much attention by the medical community and fire service. Which of the following statements correctly explains why CO has recently become such a problem?**

   A.   With the increase in the price of oil in the 1970s, today we build our homes tighter and more energy efficient.

   B.   To save fuel, we build our homes tight, and this results in a lack of fresh air infiltration. The result is an accumulation of CO and other pollutants in the home.

   C.   People were made sick by CO and some died, but often the deaths and illnesses weren't attributed to CO.

6. **Until recently, there wasn't much that could be done about the CO problem. In fact, it wasn't even recognized as a problem. Which of the following statements correctly describes why nothing was done about the CO problem?**

   A.   Deaths and illnesses caused by CO were not attributed to CO.

   B.   CO is colorless, odorless, and tasteless. People who were made sick by it, even in danger of dying from it, were not aware of the problem.

   C.   Because people were not aware of it, they could do nothing to protect themselves from it, nor would they remove themselves from the danger.

7. **CO poisoning is the leading cause of accidental poisoning death in the United States. True or false?**

8. **The home CO detector created by Dr. Goldstein remains the only effective home CO detector available to day. True or false?**

9. The author states that CO investigations are often done in a cursory manner if not incorrectly and he points out that three things are required for a fire department to make a good CO examination. What are they?

Answers

1. These are the organizations:
   - Manufacturers of CO alarms
   - Manufacturers of gas detection instruments
   - Gas utilities, propane suppliers, and oil companies
   - Appliance manufacturers
   - The medical community
   - Fire departments

   See book pg 169

2. True.

   See book pg 170

3. These are the groups:
   - Environmental Protection Agency (EPA)
   - Occupational Safety and Health Administration (OSHA)
   - Consumer Products Safety Commission (CPSC)
   - Underwriters Laboratories (UL)
   - International Approval Services (IAS)
   - American Gas Association (AGA)
   - Canadian Gas Associations (CGA)

   See book pg 170

4. IAS Requirement no. 696.

   See book pg 170

5. All are true.

   See book pg 171

6. A, B, and C are all correct.

   See book pg 171

7. True.
   ➲ *See book pg 171*

8. False. Different types of detectors have become available to compete against it in the home detector market.
   ➲ *See book pg 172*

9. These are required:
   • The right attitude
   • The essential knowledge
   • The required tools
   ➲ *See book pg 172*

The Medical Aspects of Carbon Monoxide

10

Questions

1. As we inhale oxygen, it transfers from the lungs to the blood where it attaches to the ___?___, which in turn transports the oxygen throughout the body. Fill in the blank.

2. Which of the following statements about the binding of CO to the hemoglobin molecule is *false*?

   A. A hemoglobin molecule has two binding sites for oxygen.

   B. If CO attaches to a binding site on the oxygen molecule, it acts in two ways to thwart the cellular respiration process.

   C. CO takes up space on the hemoglobin that should be reserved for oxygen.

   D. CO prevents the oxygen present in the blood from being released by the hemoglobin.

   E. This dual effect results in *tissue hypoxia* or oxygen starvation of the cells.

3. Hemoglobin has an affinity that is between ___A___ and ___B___ times greater for CO than it is for oxygen. Fill in the blanks.

4. Because of hemoglobin's affinity for CO, even small amounts of CO can be fatal. True or false?

5. Which of the following statements about CO's medical affects are accurate?

   A. As one breathes CO and the body is deprived of life-giving oxygen, the body tries to compensate by increasing its breathing rate.

   B. Answer A results in a victim's CO intake increasing. As a result, his body responds by increasing its breathing rate even more.

   C. About 25% of CO is absorbed not into the blood but into body tissue.

   D. The CO absorbed into body tissue is not held as long as CO absorbed into the blood.

   E. The CO absorbed into body tissue can result in CO being released into the victim's system after the victim is treated, tested, and shown to be free of CO, resulting in a rebound effect of CO poisoning.

6. True or false? Two people exposed to the same levels of CO for the same amount of time can exhibit different symptoms.

7. Duration of exposure is one of the factors that will determine the effect that CO has on its victims. The longer a person is exposed to CO, the higher the person's COHb level will rise. True or false?

8. Pick out the *false* statement.

   A. The reduced flow of oxygen to the heart because of CO exposure causes the deterioration and death of heart tissue.

   B. Long-term exposure to low-level CO does not contribute to lasting brain damage.

   C. COHb levels as low as 2.5% COHb have been shown to aggravate symptoms in angina pectoris patients.

9. **Another pertinent aspect of CO is its half-life. Which of the following statements about half-life are accurate?**

   A. As we breathe CO in, we also exhale it. The problem is that we give off CO more slowly than we take it in.

   B. Statement A causes retention of CO in the system, which can lead to an accumulation of CO in the blood until equilibrium is reached with ambient CO level.

   C. One half of the CO a person absorbs into the bloodstream can take as much as three hours to leave their system.

   D. A person who has a COHb level of 10% will have that level reduced to 5% after 3 hours.

10. **As a victim breathes fresh air, his COHb level is reduced. How can we speed up this process?**

11. **Which statements are true?**

    A. In the presence of pure oxygen, the half-life of CO in the victim's system is 80 min.

    B. In the presence of hyperbaric oxygen, the half-life drops to 20 min.

    C. The half-life of CO in atmospheric oxygen is 5 hours.

    D. The more CO that is in the air you breathe, the higher and faster your COHb level will rise.

    E. Altitude is a factor in COHb levels: 3% COHb at 1,400 ft is equivalent to 20% at sea level.

12. **Which of the following statements are correct?**

    A. Even doing light work in a CO-contaminated atmosphere can greatly increase your uptake of CO.

    B. Doing light work in a CO-contaminated atmosphere will not greatly increase your uptake of CO.

    C. Young children and small pets will be affected sooner by CO poisoning and more severely than normal, healthy adults.

    D. Tissues that have the greatest need for oxygen, such as the heart wall muscle and the brain, are the most quickly and adversely affected.

13. Anyone with a diminished oxygen intake is more susceptible to CO poisoning. This group of susceptible persons includes which of the following?

   A. Those with an existing heart or lung condition.

   B. The elderly.

   C. Small children.

   D. The developing fetus.

14. Why are small children more susceptible to CO poisoning than are normal healthy adults?

15. The developing fetus is more susceptible to CO poisoning than are normal healthy adults. Give four reasons for this.

16. As the mother's CO level rises, the level of CO in the fetus rises, but more slowly. If the mother is removed from the exposure and allowed to breathe clean air, her COHb level and the COHb level of the fetus will start to drop. True or false?

17. The half-life of fetal COHb is three to five times longer than the half-life of the maternal COHb. If the mother is fine after exposure to CO, the fetus is probably fine too. True or false?

18. A quick way to tell if a victim has been exposed to high levels of CO is to look for a cherry red coloration of the victim's skin. True or false?

19. Match the CO exposure in column A with the symptoms in column B.

Exposure	Symptoms
1. Mild exposure	a. Unconsciousness
	b. Nausea and vomiting
	c. Fast heart rate
2. Medium exposure	d. Death
	e. Intestinal discomfort and diarrhea (small children)
	f. Fatigue
3. Severe exposure	g. Severe headache
	h. Drowsiness and confusion
	i. Convulsions
	j. Headache, usually a frontal headache
	k. Flulike symptoms

20. One study found that 23.6% of patients presenting at hospitals with ____?____ symptoms actually suffered from low-level CO poisoning. Fill in the blank.

21. If someone is brought to the hospital with CO poisoning, it's a real possibility that someone else is at home in worse condition. True or False?

22. It is estimated that one-half of all the cases of CO poisoning go undetected. True or false?

23. CO is responsible for how many illnesses a year, which are serious enough to require the loss of workdays?

24. **CO poisoning is reputed to be a great imitator of other illnesses. What illnesses can CO mimic?**

25. **If a whole family comes down with flu symptoms at the same time, suspect CO poisoning. True or false?**

26. **True or false? A mild exposure to CO can cause its victim to slur speech and speak incoherently, leading you to believe the victim is under the influence of drugs or alcohol.**

27. **Which of the following statements about CO poisoning are accurate?**

    A. A severe exposure can occur as a result of a slow buildup of CO, or it can occur quickly with no warning after only a few breaths of high CO concentrations of the gas.

    B. The single most important thing that you can do for a victim of CO poisoning is to shut down the source of CO in his home as soon as possible.

    C. In fresh air, the CO level in a victim's blood will be cut in half in about five hours.

    D. The quicker you can remove CO from the victim's system, the less oxygen deprivation he will have to endure and the less tissue damage he will suffer.

28. As a first responder, you must not only recognize the symptoms of CO poisoning, you must also be able to treat its victims promptly and properly. Which of the following statements about treatment are stated *incorrectly*?

    A. If you carry oxygen, you should treat CO victims with it.

    B. With 100% oxygen and a tight fitting mask, the half-life of bloodstream CO drops to about 60 mins.

    C. Consider using hyperbaric oxygenation when a serious CO exposure has resulted in unconsciousness.

    D. In the case of pregnancy, hyperbaric oxygenation may be employed, but such treatment should be predicated on consultation with the hyperbaric facility and an obstetrician, as well as the patient, if she is capable.

29. Even hyperbaric patients can later suffer a relapse as CO is slowly released from muscle tissue. All such victims must be monitored for some time after treatment. True or false?

Topics for Drill

1. Open your Web browser and enter a search for "carbon monoxide death" and other related phrases. You will find a multitude of CO-related articles and links. Download or print some of this material to be used at drill. You can also get photos and information about the various home CO detectors in this way.

2. Discuss how a CO victim might be mistaken for a substance abuser. How might this affect your actions when called to investigate a sounding home CO alarm?

3. Contact your local hospital and find out if they have a hyperbaric chamber. If they do not, find the location of the nearest one. What actions will you need to take while at a CO response to obtain hyperbaric treatment for a victim of a severe CO exposure?

Answers

1. Hemoglobin.
   ➲ *See book pg 175*

2. A is false. A hemoglobin molecule has *four* binding sites for oxygen.
   ➲ *See book pg 175*

3. A. 200
   B. 270
   ➲ *See book pg 175*

4. True.
   ➲ *See book pg 175*

5. A, B, and E are accurate.
   C. About 15% is absorbed into body tissue.
   D. CO absorbed into body tissue is held longer than the CO absorbed into the blood.
   ➲ *See book pg 175*

6. True.
   ➲ *See book pg 176*

7. False. It will continue to rise until equilibrium is established with the ambient level of CO.
   ➲ *See book pg 176*

8. B. False. It does contribute to lasting brain damage.
   ➲ *See book pg 176*

9. A and B are accurate.
   C. It can take five hours to leave the person's system.
   D. The level will be reduced to 5% after five hours.
   ➲ *See book pg 176*

10. The speed of this process is increased by administering pure oxygen.
     ⮕ *See book pg 177*

11. All statements are true.
     ⮕ *See book pg 177*

12. A, C, and D are true.
     B. False. It can greatly increase your uptake of CO.
     ⮕ *See book pg 177*

13. All are correct.
     ⮕ *See book pg 177*

14. Small children have less body mass than adults and a faster metabolism. Also, small children, like the elderly, spend more time in the home and may be exposed to CO for longer periods than their parents and older siblings.
     ⮕ *See book pg 177*

15. These are the reasons:
     • The blood of the fetus has a greater affinity for CO than does the blood of its mother.
     • CO has a longer half-life in the fetus than it does in its mother.
     • A pregnant mother with CO in her blood delivers reduced oxygen to the fetus.
     • The fetus has a small body mass and rapid metabolism.
     ⮕ *See book pg 177*

16. False. The COHb level of the fetus will continue to rise, ultimately reaching a level as much as 15% higher than the peak reached by the mother.
     ⮕ *See book pg 178*

17. False. Even if the mother's COHb level is fine after exposure, COHb level in the fetus can remain dangerously high.
     ⮕ *See book pg 178*

18. Only about 10% of victims manifest this color change. It is more likely that the victim will be cyanotic due to oxygen deprivation.
    ⮕ *See book pg 178*

19. Here is the match.
    1. b, e, f, j, k,
    2. c, g, h
    3. a, d, i
    ⮕ *See book pg 178*

20. Flulike.
    ⮕ *See book pg 178*

21. True. Make a thorough search for other victims.
    ⮕ *See book pg 179*

22. False. Only one-third go undetected.
    ⮕ *See book pg 179*

23. 10,000.
    ⮕ *See book pg 179*

24. Flu, food poisoning, psychiatric illness, migraines, stroke, substance abuse, and heart disease.
    ⮕ *See book pg 179*

25. True. The flu usually does not affect everyone at the same time. CO poisoning can affect the entire family at the same time.
    ⮕ *See book pg 179*

26. False. A medium, not a mild exposure would have that effect.
    ⮕ *See book pg 179*

27. A, C, and D are accurate.
    B. The single most important thing that you can do for a victim of CO poisoning is to remove him from the contaminated atmosphere.
    ⮡ *See book pg 180*

28. B. The half-life of bloodstream CO drops to about 80 minutes.
    ⮡ *See book pg 180*

29. True.
    ⮡ *See book pg 181*

The Carbon Monoxide Emergency

11

Questions

1. Which of the following statements are accurate?

A. CO is slightly heavier than air, having a vapor density of 1.1.

B. CO is colorless, odorless, tasteless, toxic, and explosive.

C. The explosive range of CO is 5–15% in air and is rapidly fatal at 20% in air.

2. Which of the following are stated *incorrectly*?

A. As a log in a fireplace burns down to glowing embers, it gives off more CO than it did when the fire was roaring in the hearth.

B. Gas-operated refrigerators might be found anywhere that you find people living without electricity or in the home of an elderly person who didn't upgrade to an electric model.

C. A properly maintained and adjusted appliance will produce no CO.

D. An improperly adjusted flame in a furnace or on the stovetop can result in a hotter flame and thus increase the production of CO.

E. Rust flakes on a gas hot water heater's burner will raise the temperature of the gas flame, resulting in the production of excess CO.

3. Using an unvented space heater in a sealed home results in a decrease in the home's oxygen level and an increase in CO production. Some space heaters use an oxygen depletion sensor, which will shut the appliance down if it senses low ambient oxygen, thus preventing the formation of CO. This sensor will adequately protect the occupant from deadly CO. True or false?

4. All unvented fuel-burning appliances spill some CO into the home. If improperly adjusted, maintained, and operated, the amount of CO they produce will increase. True or false?

5. Which of the following are stated correctly?

   A. A properly constructed and maintained flue will vent combustion gases including CO to the outdoors.

   B. If the flue is blocked, it will spill CO into the home.

   C. The longer the flue, the better the draft and the less chance there is for CO to spill into the home.

   D. If the flue pipe is undersized, combustion gases will spill into the home.

   E. If the flue has suffered rust damage, it may perforate, and as a result, leak flue gases into the home.

6. The author uses the term *downdraft* to describe a phenomenon that is referred to as *backdraft* in articles written outside of the fire service. Which of the following statements about this phenomenon are true?

   A. Ideally, a home should be operating at neutral pressure.

   B. If the home is tightly sealed and if air is pulled or pushed out of it, a condition of positive pressure can be created inside.

   C. If negative pressure in a home is strong enough, it can result in flue gases being sucked back down and into the home.

D. A strong wind blowing across an unused chimney can cause negative pressure in the home by crating a Bernoulli effect, which can suck air up and out of the unused chimney.

E. Air can also be drawn out of a house as wind blows around the house if there are windows open on the windward side.

F. If the negative pressure created is stronger than the natural draft of your appliances, CO will spill back into the home.

G. Reverse stacking is a term used for a mechanically induced downdraft.

7. **Today's smaller, more efficient heating units are designed to prevent the escape of excess heat up the flue and instead use it to heat today's homes more efficiently. Which of the following statements about these heating units are *false*?**

A. Heat that goes up the flue is partially responsible for the draft that sucks waste gases out of the home.

B. As the temperature of the flue gas is reduced, the flue becomes less susceptible to downdrafting and reverse stacking.

C. A high efficiency furnace can become so efficient in removing heat from flue gases that it will need a fan to induce a draft in the flue.

D. The cooler the flue gas temperature, the less chance that acidic flue gases will condense on and perforate the metal flue, causing CO to spill into the home.

E. In arctic climates, condensate that seeps into the flue liner can freeze and eventually block off the flue.

8. **Warming up a car in the garage that is attached to your home can cause CO detectors in the home to go off unless the exterior garage door has been left open during the warm-up period. True or false?**

9. **Sources of CO outside the home rarely produce CO problems in the home. True or false?**

10. **Which of the following statements are true?**

   A. Home CO alarms are designed to sound before the levels of the gas become deadly, so it is unlikely that your response will require immediate emergency action in most cases.

   B. With training, your dispatcher can decide whether to send a full response plus an ambulance or a single unit equipped with a CO meter. The responding fire officer can call for additional help or reduce the response once he is on the scene.

   C. Your dispatcher can positively determine if a CO response is an emergency response or not.

   D. When the dispatcher is in doubt as to whether a CO response is an emergency or not, he should send a non-emergency response since most CO responses are non-emergency.

11. **You are a company officer in a ladder company. Your CO meter is out of service and you are dispatched to a reported CO incident. What should you do?**

12. **A meter that is left in the home for several days to measure CO levels and can then be used to produce graphs of the gas levels and times of occurrence in the home is called what?**

13. **CO poisoning has received so much publicity that doctors no longer mistake CO poisoning for the flu. True or false?**

14. **Your goals at a CO incident should include which of the following actions?**

   A. Locate the source.

   B. Shut down the source.

   C. Ventilate the area.

   D. Give immediate treatment to any victims.

   E. All of the above.

15. Once the source has been located and shut down, you should explain what three things to the occupant?

16. You have located the source of CO in the home and shut it down. Why should you take CO readings after ventilating and before you leave?

17. Which of the following statements are correct?

   A. OSHA is concerned with CO exposure in the home.

   B. The EPA is concerned with CO exposure in the workplace.

   C. The upper acceptable level for exposure in the workplace is 50 ppm for 8 hours while the level for CO exposure in the home is 9 ppm for 24 hrs. The reason the home level is lower is that exposure in the home is expected to be for a longer period than exposure in the workplace.

   D. It is expected that workers only spend eight hours at their workplace so the acceptable level of CO is higher than it is in the home.

18. Fire departments around the country have adopted different parameters for SCBA use. When deciding at what CO level to use SCBA at CO investigations, you must pick a realistic number. If the number is too high, it may be ignored by units in the field. True or false?

19. Pick the correct statements.

   A. Firefighters should conduct CO investigations in teams.

   B. Firefighters should wear their SCBA face piece for all CO investigations.

   C. When conducting a CO investigation, have a team of SCBA-equipped firefighters standing by to assist the investigation team should they run into trouble.

   D. If you find a victim that has died of CO poisoning, you should assume that all victims in the area will be dead.

20. **Fill in the blanks:**

Effects of CO exposure by ppm, time, and % in air.

ppm	time	% in air	symptoms
_A_	120 min	.02	Flu-like
800	_B_	.08	Flu-like
800	180 min	_C_	Death
1,600	_D_	.16	Death
3,200	10 min	.32	_E_
_F_	30 min	.32	Death
12,800	_G_	1.28	Death

21. **Which of the following statements are *incorrect*?**

A. A classic symptom of CO poisoning in the home is an occupant who feels sick when he is away from home and feels better when he is back home.

B. It is not necessary to check all of the occupants for symptoms.

C. If symptoms are evident, remove the occupants from the contaminated area and treat them with oxygen.

D. Start your investigation with an interview of the occupants. Ask if anyone is feeling symptoms.

22. **Which of the following statements are true?**

A. CO alarms increase during the holidays because of the accompanying prolonged use of gas ranges to cook holiday meals contribute excess CO into the home.

B. It is not uncommon for an appliance to give off excess CO as it warms up.

C. If a CO detector is sounding and you find CO in the home but you find no source of CO inside the occupancy, you should ventilate and conclude your investigation.

D. If, when you respond to a home where an alarm has sounded but find no CO present in the home, assume that the alarm is faulty.

23. **Which of the following statements about finding the source of CO in a building is stated *incorrectly*?**

   A. If you don't find a source of CO in the building, step back and get a look at the big picture.

   B. The International Association of Fire Chiefs (IAFC) in a CO update stated that CO levels in closets and cabinets will drop as quickly as the levels in the rest of the house drops once it is vented.

   C. Even if CO is present in the home, you may not be able to locate the source.

   D. If the heating system's flue is cracked, the level of CO in the home may decrease as the temperature outside drops.

24. **Which of the following statements about CO investigations are accurate?**

   A. If after a thorough search of the house for CO and defective appliances, you find nothing, there is more that you can do. You should set up a worst-case scenario.

   B. To set up a worst-case scenario, turn off all of the fuel burning appliances and fans in the home and open all of the windows and doors.

   C. The presence of stuffy, stale, or smelly air is a definite indicator of CO in the home.

   D. Certain plastic vent pipes have been recalled by the CPSC because they crack or separate at the joints, leaking CO into the home. They can be identified only by their red color.

25. **Which of the following statements are true?**

   A. A draft hood is found on oil burners.

   B. Soot around the edges of the draft hood indicates the spillage of flue gases.

   C. Perform a draft test by lighting an incense stick or match and holding it near the hood while the burner is on.

   D. Answer C and if the smoke is drawn up into the flue, there may be a CO spillage problem.

26. **If you do find high levels of CO in a home, shut down the offending appliance. If you can't find the source, which of the following should you do?**

A. Shut down all of the appliances.

B. Evacuate the home.

C. Ventilate the home.

D. Retest the home and when it is safe, allow the occupants to return.

E. Instruct the occupants not to operate their appliances until they have been properly checked by the utility or repair service.

Topics for Drill

1. CO sources.

   A. Survey your area for locations where outdoor sources of CO could cause a CO problem within a structure.

   B. Make a list of all of the possible external sources.

   - Check residences as well as commercial occupancies and places of public assembly. Your sources should include power tools, vehicles, and outdoor grills.

   - Review this list with your firefighters at drill and ask them to think of other possible sources.

   - Make a checklist that contains all of the sources that you have discovered. Leave room to add new sources as you discover them. Do a similar survey in various types of buildings to locate interior sources of CO. Again, make a checklist and periodically review it with your firefighters.

2. Analyzing information.

   A. Keep a record of the levels of CO that you find when responding to CO alarm activations.

   B. Make a chart of these levels and the effect they could have on occupants after several different time spans.

   C. Use this chart to keep your firefighters mindful of the effects of short-term exposure and long-term exposure to CO.

Scenario: Carbon Monoxide Emergency

You are a lieutenant in a truck company. Your unit is staffed with an officer and four firefighters. It has been a cold winter and tonight is no exception. The temperature is in the single digits. It is 2400 hours and you have just received a call for a CO alarm sounding in a two-family house. You are dispatched alone to this alarm. When you arrive, you find that the reported address is a two-story woodframe building attached to a block of similar buildings. The reported address is in the middle of the block. There is a driveway that serves all of the buildings, running behind the row of dwellings.

These buildings were built with garages in the rear, but some people have converted them to illegal apartments. You go up to the front door and knock on the door to the first floor apartment. A man answers the door and explains that his wife told him that his CO detector has been sounding on and off for the last two days and he is worried. He is an over-the-road trucker and he just got home a few hours ago. His wife and infant are in the back room asleep. He asks you to be quiet as they have the flu and just got to sleep. You are equipped with a CO meter.

Scenario Questions

1. What will you do?

   A. Where should you start taking readings with your CO meter?

   B. How would you conduct your survey inside with your CO meter?

   C. At what level of CO would you evacuate the building?

D.   What questions should you ask the occupant?

E.   What help if any would you call at this time?

F.   What immediate action would you take to safeguard his family?

G.   What must you consider before evacuating the family? Remember it is winter.

2.   You find 80 ppm CO in the 1st-floor apartment, but do not find a source of CO in the apartment.

A.   Describe how you would expand your search.

B.   There is no answer when you ring the bell for the upstairs apartment. The occupant of the 1st floor tells you that he has heard no one upstairs all night. Do you force entry into the upstairs apartment and if so why?

C.   What if you do not find a source of CO in the building?

D.   Do you need any additional units at this time? If so, what help do you need?

Answers

1. B.
   A. Lighter—0.96.
   C. Explosive range is 12.75–74%, fatal at 1.28%
   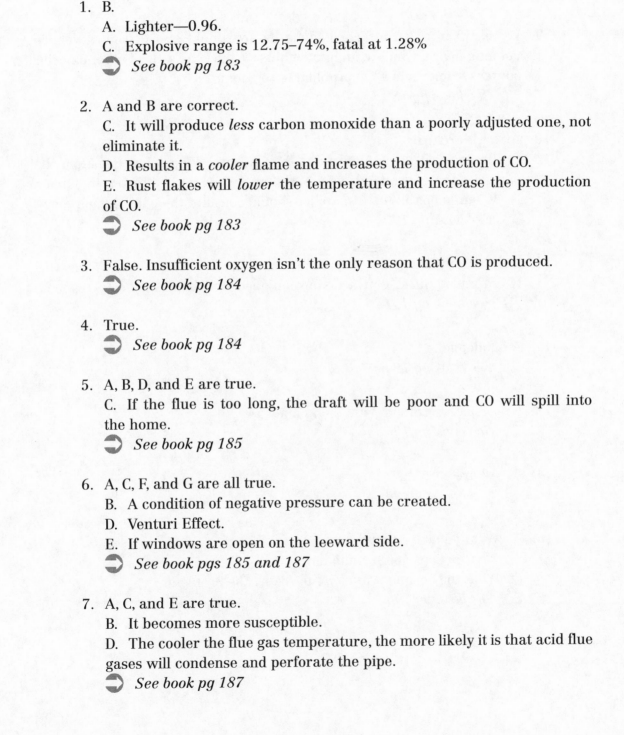 *See book pg 183*

2. A and B are correct.
   C. It will produce *less* carbon monoxide than a poorly adjusted one, not eliminate it.
   D. Results in a *cooler* flame and increases the production of CO.
   E. Rust flakes will *lower* the temperature and increase the production of CO.
   See book pg 183

3. False. Insufficient oxygen isn't the only reason that CO is produced.
   See book pg 184

4. True.
   See book pg 184

5. A, B, D, and E are true.
   C. If the flue is too long, the draft will be poor and CO will spill into the home.
   See book pg 185

6. A, C, F, and G are all true.
   B. A condition of negative pressure can be created.
   D. Venturi Effect.
   E. If windows are open on the leeward side.
   See book pgs 185 and 187

7. A, C, and E are true.
   B. It becomes more susceptible.
   D. The cooler the flue gas temperature, the more likely it is that acid flue gases will condense and perforate the pipe.
   See book pg 187

8. False. It can cause the CO detectors to go off even if the exterior garage door is left open.

  ➲ *See book pg 188*

9. False. CO from an attached home or apartment or CO from a vehicle running near a window, an intake fan, or a duct , as well as other exterior sources, can create a CO problem in the home.

  ➲ *See book pg 189*

10. A and B are true.

  C. He can't positively determine this, but by asking a few questions, he can get a pretty good idea as to whether anyone in the home is in danger.

  D. When in doubt, the dispatcher should consider the call an emergency.

  ➲ *See book pg 190*

11. Immediately notify the dispatcher that your CO meter is out of service (OOS) and request that a meter-equipped unit be assigned to respond.

  ➲ *See book pg 191*

12. A datalogger.

  ➲ *See book pg 194*

13. False. Such errors are still being made.

  ➲ *See book pg 196*

14. E. All are correct.

  ➲ *See book pg 196*

15. A. What you have found.

  B. What actions you have taken.

  C. What the occupant must do to rectify the problem.

  ➲ *See book pg 196*

16. To ensure that you haven't missed another source of CO.
    ➥ *See book pg 196*

17. C and D are correct.
    A. OSHA is concerned with exposure at the workplace.
    B. The EPA is concerned with exposure in the home.
    ➥ *See book pg 197*

18. False. If too low, it may be ignored by the units in the field.
    ➥ *See book pg 197*

19. A and C are correct.
    B. Firefighters should wear their face pieces in the standby position and put it on when a predetermined level is reached.
    D. The effects of CO on individuals can vary.
    ➥ *See book pg 198*

20. A. 200 ppm
    B. 45 min
    C. .08%
    D. 60 min
    E. Flu-like
    F. 3,200
    G. 1–3 min
    ➥ *See book pg 198*

21. C and D are correct.
    A. A classic symptom of CO poisoning in the home is feeling sick when at home and better when away from home.
    B. It is necessary to check all occupants for symptoms.
    ➥ *See book pg 199*

22. A and B are true.

    C. False. If you find no source in the occupancy, expand your search. The source may be coming from outside of the occupancy.

    D. Consider the possibility that there is an intermittent source that has now dissipated or that the occupant has ventilated the home prior to your arrival.

    ⟳ *See book pgs 199 and 200*

23. A and C are true.

    B. The levels in closets and cabinets may remain high even after the home has been ventilated. This will occur after a high level of CO has built up in the home.

    D. As the outside temperature drops, the furnace must work harder to heat the home. As a result, it pumps more CO into the home through the cracked flue. The level of CO will increase.

    ⟳ *See book pg 201*

24. A is true.

    B. False. Turn on all appliances and close all of the windows and doors.

    C. False. CO is odorless. Stuffy, stale, or smelly air indicates that the house is tightly sealed with little air exchange.

    D. False.
    1. They are grey or black in color.
    2. They pass through the house wall not through the roof.
    3. The names *Plexvent*, *Plexvent 2*, or *Ultravent* are stamped on the vent pipe.

    ⟳ *See book pgs 202–204*

25. B and C are true.

    A. It is found on gas burners.

    D. If the smoke is *not* drawn up into the flue, there may be a spillage problem.

    ⟳ *See book pg 204*

26. All answers are correct.

    ⟳ *See book pg 207*

Scenario Answers

1. A. Start taking readings outside of the building to determine the ambient level of CO. If there is an inversion, ambient levels alone might be high enough to set off the alarm.

   B. Take readings at the door to the apartment then take readings throughout the apartment at various levels. Check near the stove and any other fuel-burning appliances. Check near the heat registers if it is a hot air heating system. (Remember, the hot water heater, heating unit, and stove must be running and warmed up in order for you to get accurate readings.)

   C. Fire Department of New York (FDNY) would require mandatory evacuation if the level were above 100 ppm. You should have a set evacuation level for your department. If you do not have one, now is a good time to set one. Check with nearby departments and see what levels they have set.

   D. Ask the occupant what was going on in the home when the alarm sounded. If the shower was in use, it might indicate that the hot water heater was at fault; if the stove was in use it might implicate the stove. Did anyone warm a car up in an attached garage? Ask where the alarm is located and what type of alarm it is. If it is in the kitchen or near heating equipment, it might be triggered unnecessarily. Your knowledge of home CO detectors should give you an idea of the levels required to trigger the alarm. Ask how long his family has been sick and if they both got sick at the same time. This might be a case of CO poisoning and not the flu. The truck driver has been away and will not have been exposed to the CO for a prolonged period of time so he might not yet be feeling the symptoms.

   E. Call for an ambulance to check out his family. It will be necessary to wake them to see if they are OK.

   F. If his family is suffering from the effects of CO, you should remove them from the contaminated atmosphere. This can be done by taking them outside of the home, and if necessary, giving them oxygen.

G. Where are you going to put the family once you remove them from the building? It is cold outside. You must provide shelter for them. Consider putting them in a neighbor's home or in an apparatus to keep them warm. Will they have to be relocated? What system do you have in place for such an event?

2. A. The source may well be coming from the heating unit that is probably located in the basement. You must expand your search to all of the fuel-burning appliances in the building. Don't forget the car in the garage and any possible illegal apartment.

B. You must get into the upstairs apartment. The occupants may be sleeping, they may be unconscious from the CO, or they may be out. You must check to verify that they are not suffering from CO poisoning.

C. If you do not find a source of CO in the building, check the adjoining buildings on either side. A cracked flue in a common chimney could be pumping CO into the two-family dwelling. This will necessitate waking up the occupants on either side of the original building and even forcing entry if no one answers the door. Consider outside sources of CO, such as vehicles, tools, open fires, and any other possible source. Remember, the alarm was sounding for several days, off and on.

D. Since you are now checking several buildings, you will need more firefighters and at least one more CO meter. More meters would be better. If you have not done so already, you should call a chief officer to the scene to manage the overall operation. If you have to evacuate more than one building, you must provide shelter for all evacuated. Consider calling a bus to serve as a temporary shelter. If you have forced entry into any of the apartments or structures, you may need the police to provide security for the now-open residences.

12

Home CO Detectors

Questions

1. **Which of the following statements about the biomimetic CO detector are true?**

   A. Biomimetic detectors are based on synthetic hemoglobin contained in a small, round disk or gel cell.

   B. This gel cell mimics the body's hemoglobin by absorbing ambient CO in much the same way our own hemoglobin absorbs it.

   C. As the gel cell absorbs CO, it begins to change color.

   D. The gel cell is black, but in the presence of CO, it changes to a milky white as it absorbs CO.

2. **Mark the following statements about biomimetic detectors as *true* or *false*.**

   A. The gel cell's color changes are monitored by a light-emitting diode that triggers the alarm once a predetermined level of CO is reached.

   B. The more CO that the gel cell absorbs, the lighter it becomes.

   C. A black coloration indicates that the cell has been killed and will take time to reset.

   D. If the gel cell is white, it may indicate that it was exposed to heavy concentrations of carbon monoxide.

3. **Which of the following are stated correctly?**

    A. As a gel cell absorbs CO, it also releases CO and it releases the gas faster than it absorbs it.

    B. To purge the early model gel cell sensor of CO gas, place it in uncontaminated air for 2–24 hrs.

    C. If the early model gel cell detector takes more than 48 hours to clear, it indicates that it has been killed by CO and must be replaced.

    D. The time required to clear the sensor depends on the amount of CO the detector has been exposed to.

4. **How would you stop the alarm from sounding in an early model biomimetic detector?**

5. **The early biomimetic detectors did not offer continuous protection to the occupant once the alarm sounded. True or false?**

6. **The new model biomimetic detectors offer continuous protection to the occupant, even after the alarm has triggered because of a CO buildup in the home. Which of the following about these newer detectors are stated *incorrectly*?**

    A. Once it alarms, it will continue to sound until the CO level drops below the trigger point.

    B. The newer model has a reset button that will silence the alarm for 6 minutes even if the CO levels remain high.

    C. If after pressing the silence button, the ambient CO levels drop sufficiently, the detector will go into its sentry mode and again be ready to warn of a new buildup.

    D. If the levels do not drop after pressing the silence button, the alarm will sound again after 10 minutes.

7. **The following is a listing of features of the gel cell detector. Mark them _O_ for the older version or _N_ for the newer version.**

   A. A white detector with the words Carbon Monoxide Detector printed in white on the face.

   B. A white detector with the words Carbon Monoxide Detector printed in black on the face.

   C. A detector that has the word FACOR printed on the side of the gel cell/battery module.

   D. A detector that has the word NICOR printed on the side of the gel cell/battery module.

   E. A sensor module that is black.

   F. A sensor module that is white.

8. **Which of the following are true statements about gel cell detectors?**

   A. Both the old and the new type sound full alarm with a continuous horn and warn of low battery with an intermittent chirp each minute.

   B. Both the old and the new alarms have an early warning alarm of three to five chirps every five minutes.

9. **Knowing the various detectors and the types of alarms they can sound will help you in your investigation. Pick out the _incorrect_ statements.**

   A. Asking the homeowner to describe the sound the alarm made may help you determine the reason it sounded and the potential danger posed to the occupant.

   B. If the battery alert sounded on a gel cell detector, you can advise the occupant to replace the module.

   C. If the early warning alarm sounded, you'll know that the level of CO in the house didn't reach dangerous levels.

   D. If the full alarm sounded, you may have a dangerous situation.

10. **Match column A with column B.**

A	B
1. Full alarm	a. Chirp each minute
2. Low battery	b. 3–5 chirps each five minutes
3. Early Warning	c. Continuous
4. Replace unit battery/module	d. 2–3 years

(For biomimetic detector)

11. **First Alert has come out with a new portable biomimetic CO alarm. Which of the following *does not accurately* describe this device?**

A.   It has a triangular shape with rounded ends.

B.   It has a replaceable 9V battery and a replaceable CO sensor.

C.   It must be wall mounted.

D.   It is easily carried in a suitcase and may be brought into motels by safety-conscious guests.

12. **The new portable biomimetic CO detector sounds slightly differently alarms than the earlier models. It also indicates the type of alarm by sound and color of a flashing light. Match column B and C with column A.**

A	B	C
Warning	Type of Alarm Sound	Flashing Light Color
1. Full alarm	a. Chirp 2X/min	x. Flashing red light
2. Low battery	b. Chirp 3X/30 sec	y. Flashing yellow
3. Replace Unit	c. Continuous	z. Flashing green

13. **Still another new type biomimetic detector is the combination smoke/CO detector. Which of the following statements about it are *incorrect*?**

   A. It has a pentagonal shape.

   B. It contains both photoelectric smoke detector and a biomimetic sensor module that has been designed to resist low-level alarms.

   C. It has a single test/silence button and uses two replaceable C cell batteries.

   D. Pressing the silence button will silence the smoke alarm for 8 mins, whereas the CO alarm is silenced for 4 mins.

   E. Both functions of the unit smoke detector and CO detector have a distinctive alarm accompanied by an identifying flashing light.

14. **For the combination smoke/CO alarm, match the alarm with the sound and warning light.**

Alarm	Sound	Light
1. Smoke	a. Single on-off tone, each 1½ sec long	A. Yellow light
2. CO	b. Three loud, consecutive beeps, then a pause	B. Flashing red dot pattern
3. Low battery	c. Rapid chirp	C. Flashing indicator light
4. Replace unit	d. Warning chirp	D. Flashing red flame-shaped light

15. **Another type of home CO detector—the semiconductor detector—does not contain a gel cell. Which of the following statements about the semiconductor detector are correct?**

   A. It uses a small ceramic component coated with tin dioxide powder to detect CO.

   B. An embedded heating element periodically burns off moisture, CO, and oxygen as well as other contaminants.

   C. The heating element is connected to an integrated circuit that monitors the sensor.

   D. Oxygen decreases the electrical resistance of the wire while CO increases it, allowing electrons to flow more easily.

   E. As CO accumulates on the sensor, the lowering of the electrical resistance is noted by the microprocessor; at a predetermined level of CO, the detector alarms.

16. **The semiconductor alarm burns contaminants off of its sensor every 2½ mins and actually resets itself each time it starts a new cycle. True or false?**

17. **As semiconductor detectors age, they become less sensitive. True or false?**

18. **The IAFC has warned that semiconductor detectors might be poisoned by something that is commonly found in many homes. What was the poisoning agent and how would the detector react to it?**

19. **How many substances can you list that, if present in high concentrations, can temporarily cause a semiconductor detector to incorrectly sense CO.**

20. List three substances that can permanently poison a semiconductor detector.

21. We currently have gel cell detectors, semiconductor detectors, and, more recently, a third technology is being used in the manufacture of home CO detectors. What is the third type of CO detector?

22. The electrochemical detector is sensitive to certain gases that may affect the reliability of the unit. What are these gases?

23. The AIM electrochemical detector stores the peak CO level and peak COHb level for the prior 24 hours. How can you access this data?

24. For what purpose is the green home detector offered by AIM intended to be used?

Topics for Drill

1. Collect the instruction booklets from as many different types of home CO alarms as you can and review the features offered by each type of alarm. From the information contained in the booklets and from *Responding to "Routine" Emergencies*, make a chart that will show the following:

    a. How the various alarms differ.
    b. The levels at which each alarm can be expected to sound.

2. Find the contact number for the alarm company on the booklets that you have collected. Call the company and tell them that your fire department would like any educational material they can send you on CO dangers and how the fire department should conduct CO investigations. Tell them that you are training your firefighters and would like a sample of one of their detectors to use in this training.

3. Take the detectors that you received from the alarm company or that you purchased and mount them on a board. Have your firefighters describe the characteristics of each detector on the board. If you do not have detectors, use photos or promotional pictures received from the alarm companies. You can use this board in public education campaigns.

Answers

1. A, B, and C are correct.

   D. The color starts out as translucent orange, but in the presence of CO, it changes to a darker orange and then an olive green. As the amount of CO increases, the color then changes through brown and black.

   ⮑ *See book pg 209*

2. A. True

   B. False. The darker it becomes.

   C. False. It will not regenerate, the sensor must be replaced.

   D. False. White may indicate exposure to heavy concentrations of steam.

   ⮑ *See book pg 209*

3. B and C are correct.

   A. It absorbs gas faster than it releases it.

   D. It depends on both the amount of CO it has been exposed to and for how long it was exposed.

   ⮑ *See book pg 209*

4. Remove the battery/sensor module.

   ⮑ *See book pg 210*

5. True. Since the battery/sensor module had to be removed and placed in clean air to purge, it no longer was available to warn of a new or continued CO gas buildup.

   ⮑ *See book pg 210*

6. A, B, and C are correct.

   D. The alarm will sound again after 6 min.

   ⮑ *See book pg 210*

7. A, C, and E are features of the older detector.

   B, D, and F are features of the newer detector.

   ⮑ *See book pg 211*

8. A is true.

    B is false. Only the new type has the warning alarm. [First Revision of UL 2034 (1995) The third revision of UL 2034 (1998) does not allow a warning alarm.] *Note: The third revision occurred after the printing of the text and is not mentioned. For more information on this topic, go to the author's website at http://chiefmontagna.com and click on the link "book updates."*

    ➲ *See book pg 213*

9. A, B, C, and D are correct.
    ➲ *See book pg 213*

10. 1-c, 2-a, 3-b, 4-d.
    ➲ *See book pg 213*

11. D is correct.
    A. It is rectangular with rounded edges.
    B. The CO sensor is not replaceable
    C. It can be placed on a dresser, table, or countertop or can be wall mounted.
    ➲ *See book pg 213*

12. 1, c, x
    2, a, y
    3, b, y (green is not mentioned)
    ➲ *See book pg 214*

13. D and E are correct.
    A. It has an octagonal shape.
    B. It contains an ionization smoke detector and biomimetic sensor module.
    C. It uses a single replaceable 9V battery.
    ➲ *See book pg 214*

14. 1, b, D
    2, a, B
    3, d, C

4, c, A
⟳ *See book pg 215*

15. A, B, C, and E are correct.
   D. Oxygen increases the electrical resistance, while CO decreases it.
   ⟳ *See book pg 215*

16. True.
   ⟳ *See book pg 215*

17. False. They become more sensitive and sound their alarms at lower ambient levels of CO.
   ⟳ *See book pg 216*

18. High levels of natural gas (10,000 ppm) caused the semiconductor home detector to alarm as if exposed to CO.
   ⟳ *See book pg 217*

19. Methane, propane, isobutane, ethylene, ethanol, alcohol, isopropanol, benzene, toluene, ethyl acetate, hydrogen, hydrogen sulfide, sulfur dioxides, aerosol sprays, alcohol based products, paints, thinners, solvents, adhesives, hair sprays, aftershaves, perfumes, and some cleaning agents.
   ⟳ *See book pg 217*

20. Sulfur dioxide, hydrogen sulfide, and silicone.
   ⟳ *See book pg 217*

21. Electrochemical detectors.
   ⟳ *See book pg 217*

22. Hydrogen, ethanol, sulfur dioxide, and hydrogen sulfide.
   ⟳ *See book pg 218*

23. You must have the appropriate AIM gas detection meter, the AIM Safe Air Reader.

➲ *See book pg 219*

24. It is intended for use by heating, ventilation, and air conditioning (HVAC) technicians and fire departments. It is to be left overnight in a suspect home then picked up the next day and analyzed. The green color makes it easily identifiable as fire department property.

➲ *See book pg 219*

UL 2034

Note from author: The following information reflects a revision in UL 2034 that occurred on October 1, 1998. Contact Underwriters Labs (UL) for complete information. Log on to my Web site at http://chiefmontagna.com and click on the link "book updates" for a brief breakdown of the revisions.

UL 2034 — Quick Review and Recent Revisions

- UL 2034 was first published in April 1992.

    - Revised October 1995

    - Revised again October 1998

- Both battery and house current are allowed.

- Alarm should sound when COHb in a healthy adult would reach 10% COHb.

    - 10% reached after 90 mins exposure to 100 ppm

- The 1995 standard required that the minimum and maximum levels for alarms be stated.

- Low-level warning signal is not required on 1992 standard. It is permitted but not required on the 1995 standard. *(The new 1998 standard does not permit a warning signal.)*

- UL – required response times

Required 1995 UL 2034 level	Required 1998 UL 2034 level
100 ppm within 90 mins	*70 ppm within 189 mins (minimum 60 mins)*
200 ppm within 35 mins	*150 ppm within 50 mins (minimum 10 mins)*
400 ppm within 10 mins	*400 ppm within 15 mins (minimum 4 mins)*

- Resistance to low-level alarms

  - 1992 standard is to resist alarming at 15 ppm for 15 days *(correct number may be 8 hours)*

  - 1995 and new 1998 standards are to resist alarming at 15 ppm for 30 days

- The old 1992, 1995, and 1998 standards have two alarms: a trouble alarm and a full alarm signal. The 1995 standard allows but does not mandate the addition of a warning alarm.

- The 1995 and the new 1998 standard have a reset button.

- The 1995 and the new 1998 standard require contrasting markings.

- The new 1998 standard requires detailed instructions be placed on the detector.

- The old 1992 standard instructed occupants to call the fire department and leave the premises.

- The newer 1995 standard told occupants to leave the premises only if symptoms were present.

- The new 1998 standard again warns residents to call the fire department and to leave the premises.

- The International Approval Service (IAS), formed by gas companies, has its own standards for CO alarms.

- The Consumer Product Safety Commission (CPSC) has its own standards. (Some of these were incorporated into the new 1998 UL 2034.)

Questions

1. According to the information provided in *Responding to "Routine" Emergencies*, there are three generations of CO detectors available and in use. Which of the following statements about them are true?

   A. Those built and sold before April 1992 were manufactured to no standard.

   B. UL began to research the standard for CO detectors in 1989.

   C. UL 2034 was completed and published in April 1992 and revised in October 1999.

   D. You will encounter all three generations of detectors in the field.

2. The 1992 version of UL 2034 required that the detectors sound an alarm when the CO present in the air would cause anyone's COHb level to reach 15%. True or false?

3. At 10% COHb a healthy adult would exhibit symptoms. True or false?

4. Which of the following statements are true?

   A. For a normal, healthy adult working at a moderate rate, this 10% level will be reached in 90 min exposure to 100 ppm of CO.

   B. If the ambient CO level is 200 ppm, it will take a normal healthy adult 35 mins to reach 10% COHb level.

   C. If the ambient CO level is 400 ppm, it will take a normal healthy adult only 20 mins of exposure to reach the 10% COHb level.

5. Both the 1992 and the 1995 standards require that a CO detector resist alarming when exposed to 15 ppm of CO for 8 hours. True or false?

6. Which are the true statements regarding the 1995 revision of UL 2034?

   A. It is expected that there will be periodic, short-lived high levels of CO in the home, and the UL standard allows for these.

   B. Detectors are expected not to sound an alarm when exposed to 50 ppm for 16 min, but they are expected to sound before 90 min exposure to 100 ppm.

   C. Both the 1992 and 1995 standards require a trouble signal that gives a short chirp or beep approximately once a minute.

   D. A trouble signal means that the battery needs to be replaced or the detector needs to be replaced or serviced.

7. The occupant will probably call the fire department whenever the alarm goes off, unaware that this particular device may be capable of sounding several different types of alarms. True or false?

8. Which of the following statements are incorrect regarding warning alarms?

   A. All CO detectors issue an early warning signal, indicating a buildup of gas that has not reached the prescribed activation levels.

   B. The warning signal will sound as an intermittent alarm for approximately 3–5 sec every 3–4 min.

   C. The revised UL 2034 (1995) standard states that the alarm activation levels of a detector must be stated in the owner's manual.

9. The following is a list comparing features required by the old and the revised UL 2034 (1995). Circle the correct answer for each category under the original and revised standard.

R – Required   NR – Not Required

SA – Sounding Alarm   S – Symptoms

O – Optional

Feature	Original Standard	Revised Standard (1995)
1. Reset button	R / NR / O	R / NR / O
2. Alarm level disclosure	R / NR / O	R / NR / O
3. Contrasting lettering	R / NR / O	R / NR / O
4. When to call fire department	SA / S / O	SA / S / O
5. Rush hour test	R / NR/ O	R / NR / O
6. Early warning	R / NR / O	R / NR / O

10. What organization made up of gas companies in the United States and Canada developed and published supplemental standards for CO detectors?

11. Why does the CPSC recommend the elimination of the early warning signal on home CO detectors?

12. **Which of the following statements are true?**

    A.   The CPSC recognizes that certain areas of the country have high ambient levels of CO.

    B.   The CPSC recognizes that unvented appliances combined with high ambient CO levels can give off high levels of CO in the home and can cause detectors to sound even if the home's CO level is not an immediate safety hazard.

    C.   The CPSC standards would lessen the likelihood that a thermal inversion would cause numerous alarms to sound.

    D.   The CPSC suggests that one type of CO detector is suitable for all risk groups.

13. **Which of the following statements are *inaccurate*?**

    A.   The CPSC believed that a properly functioning unvented gas appliance can trigger a CO alarm.

    B.   The CPSC believed that the existing false-alarm resistance points are unrealistically low.

    C.   The CPSC believed that the levels set by UL translate into numerous unnecessary alarms.

    D.   The CPSC recommends that the list of different gasses that CO detectors should resist alarming be reduced.

14. **Because a properly operating gas appliance produces carbon dioxide (CO_2), the CPSC suggests that detectors should resist alarming at 4000 ppm of CO_2 rather than the 400 ppm of CO_2 currently specified.**

15. **Detectors that provide a digital readout of ambient carbon monoxide start displaying CO at varying levels and with different accuracy. What does the CPSC recommend for these detectors?**

16. For years, we have educated the public about the necessity and proper placement and maintenance of smoke detectors. We should perform the same education function for CO detectors. True or false?

17. **Which of the following statements are accurate?**

   A.   Gas-absorbing cards that change color in the presence of CO are good alternatives to home CO detectors and cost much less.

   B.   In the summer, public service announcements should be broadcast warning of CO dangers.

   C.   In the winter, power failures that accompany winter storms often result in a loss of heat and the use of gas range or space heaters for long periods of time to heat homes. This practice puts occupants at risk of CO poisoning.

   D.   If you have a large non-English speaking population, broadcast your public service announcements in their native language.

   E.   A good time to pass along information about CO is when a recent CO poisoning incident is in the news.

18. **Which statements are accurate?**

   A.   Unlike smoke alarms, CO alarms can be placed within 6 in. of where a wall meets the ceiling.

   B.   Don't place the CO detector behind curtains or furniture or anything else that would block the airflow to it nor near a HVAC supply register or fan.

   C.   Don't place a CO detector within 15 ft of an appliance.

   D.   A CO detector in the kitchen is a good idea, because it is positioned to give an early warning of a defective stove.

   E.   Don't place a CO detector in an uninsulated space.

   F.   The CPSC recommends that CO detectors be placed on each level of the home near sleeping areas.

   G.   Manufacturers recommend that a detector be placed wherever in the home people congregate.

   H.   A good time to explain to occupants where detectors should be placed is when you respond to their home for a CO alarm sounding.

19. The CPSC has published a standard for the placement of CO detectors, and it is called CPSC 900. True or false?

Topics for Drill

Prepare an educational program that informs homeowners and apartment occupants of the following:

a.   Why a CO meter is necessary.

b.   Where to place the home CO alarms.

c.   How to maintain the alarm.

d.   What they can do to prevent a CO emergency in their home.

e.   The types of alarm available.

This public service message can be delivered at public gatherings, and if you get the cooperation of the local media, on TV and/or radio. A good time to deliver this message is after a CO incident in your area, as winter approaches, and before major storms where power outages are a concern. Consider the possibility that you may need to deliver the message in the language spoken by the people living in your area.

Answers

1. A, B, and D are true.
   C. Revised in October 1995 (*Note: UL 2034 was revised again in 1998 after the publication of this book. This now gives us four generations of CO detectors.*)
   ➲ *See book pg 223*

2. False. It required the alarm to sound when the COHb in a *healthy adult* would reach *10%*.
   ➲ *See book pg 223*

3. False. No symptoms would be evident.
   ➲ *See book pg 223*

4. A and B are true.
   C. 15 min.
   ➲ *See book pg 224*

5. False.
   1992 standard—resist alarming for 8 hours
   1995 standard—resist alarming for 30 days
   ➲ *See book pg 225*

6. A, C, and D are true.
   B. Detectors should not alarm when exposed to 100 ppm for 16 min but are expected to sound before 90 min exposure to 100 ppm.
   ➲ *See book pg 225*

7. True.
   ➲ *See book pg 225*

8. B and C are correct.
   A. Some detectors have an early warning signal. [*The most recent revision of UL 2034 (1998) does not allow early warning alarms.*]
   ➜ *See book pg 226*

9. Here is how the features compare.

1.	NR	R
2.	NR	R
3.	NR	R
4.	SA	S
5.	NR	R
6.	NR	O

   ➜ *See book pg 227*

10. The IAS.
    ➜ *See book pg 227*

11. They feel that it only confuses the purchaser.
    ➜ *See book pg 229*

12. A, B, and C are true.
    D. Since certain risk groups require more stringent protection, the CPSC further suggests that detectors be developed and sold specifically for their use.
    ➜ *See book pg 230*

13. A, B, and C are accurate.
    D. The CPSC recommends that the list be expanded.
    ➜ *See book pg 230*

14. False. They should resist alarming at 5000 ppm of CO_2 rather than the current 1000 ppm.
    ➜ *See book pg 230*

15. CPSC recommends a minimum standard of accuracy for digital readouts be established.
   ➲ *See book pg 230*

16. True.
   ➲ *See book pg 231*

17. C, D, and E are accurate.
   A. The cards sound no alarm and will not wake you up when CO builds up. Get a UL-approved detector.
   B. In the winter.
   ➲ *See book pg 232*

18. B, E, F, G, and H are accurate.
   A. They should be placed like smoke alarms and shouldn't be within 6 in. of where the wall meets the ceiling.
   C. The distance depends on the manufacturer's instructions. Depending on manufacturer, the distances are 5–20 ft from an appliance. I favor the 20-ft figure.
   D. A CO detector in the kitchen is a bad idea.
   ➲ *See book pg 232*

19. The NFPA has published its standard NFPA 720.
   ➲ *See book pg 233*